Cambridge IGCSE® Physics Summarised

About the author

Kaleem Akbar was born on the outskirts of Glasgow, Scotland in 1980. He graduated with a B.Sc. Honours degree in Optoelectronics and Laser Engineering from Heriot Watt University in 2002 and went on to do an M.Sc. at The University of St Andrews, before completing his PGDE (Post Graduate Diploma in Education) in Physics at Strathclyde University in Glasgow. He taught in Scotland before moving out to the Middle East and has taught Physics at IGCSE, AS and A2 level since September 2006. He wrote this book in response to his students' thirst for the essential details to achieve their full potential.

Cambridge IGCSE® Physics Summarised

Kaleem Akbar

www.igcsephysics.com

www.fast-print.net/store.php

Cambridge IGCSE® Physics Summarised
Copyright © Kaleem Akbar 2014

ISBN: 978-178456-089-8

All rights reserved

No part of this book may be reproduced in any form by photocopying or any electronic or mechanical means, including information storage or retrieval systems, without permission in writing from both the copyright owner and the publisher of the book.

The right of Kaleem Akbar to be identified as the author of this work has been asserted by him in accordance with the Copyright, Designs and Patents Act 1988 and any subsequent amendments thereto.

A catalogue record for this book is available from the British Library

The example questions and answers contained in this book have been written by the author.

Cambridge International Examinations bears no responsibility for the example answers.

® IGCSE is the registered trademark of Cambridge International Examinations.

First published 2014 by
FASTPRINT PUBLISHING
Peterborough, England.

Acknowledgements

The author would like to thank the following professionals who have helped to make this book possible:

Lynda Anderson-Coe for editing the Physics content
Kate Jamieson for copy editing
Amanda Harman and **Jane Roth** for proof reading
Craig Walton and **David Millar** for creating the illustrations
Craig Walton for editing the illustrations and the front cover
Shahad Al Qattan for designing the front cover

A special thanks to my wife **Toni Reid** for her unflappable support and to my good friend **Alistair Rae** for his counsel.

Thanks are also due to the countless students who have provided their invaluable feedback and recommendations for improving my original notes over the years.

Important information for students

The Cambridge IGCSE Physics 0625 syllabus

The syllabus contains **Core** and **Supplement** material. Core material should be regarded as basic material that all students must know for achieving a grade C. Supplement (or extended) material is more demanding material which will allow the student to achieve a higher grade.

Features in the book

Symbols

This book uses symbols to allow you to understand which material is Core and which is Supplement.

❑ Core material
○ Supplement material

Examples

The book contains examples of the types of calculations you may have to perform. All questions and answers have been written by the author.

Top Tip

Top Tip boxes contain useful advice and guidance that will help you to work to the best of your abilities.

Note

Note boxes summarise or highlight some important ideas.

Units

The SI system of units is the world's most widely used system of measurement. SI is an abbreviation for le Système International d'unités.

The system has seven base units including the kilogram (kg), metre (m) and second (s).

There are other derived units including the metre per second (m/s), as well as recognised multipliers such as mega (M), kilo (k), centi (c) and milli (m).

All of the units used in this book are part of the SI system.

How to use this book

- You can show that you have learned the statements and explanations in each section with a tick in the checklist on page xii next to each topic or next to each square or circle bullet in the main text.
- A column has been left on the outside of each page so that you can make additional notes.
- Highlighter pen can also be used to emphasise important statements throughout this book.
- At the back of the book you will find additional support material with a list of all the formulae expressed in different formats, including the triangular format. Simply cover up the quantity you want to calculate and the triangle will show whether the other quantities have to be multiplied or divided.
- There is a glossary of terms at the back of the book. A glossary is a brief dictionary that will help you to understand any words or phrases you're not sure of.
- The glossary of examination terms will help you to understand how you should answer examination questions.

Cambridge IGCSE Physics

The course

This book has been designed to cover all of the learning outcomes for the latest Cambridge IGCSE® Physics 0625 syllabus for examinations from 2016 onwards. You should be aware that the syllabus can vary slightly from year to year.

Examination structure

All candidates are expected to take three separate examination papers as outlined below.

Candidates who have studied the Core syllabus content will take **Papers 1, 3 and either 5 or 6**. The maximum grade that can be achieved from these papers is C.

Candidates who have studied the Core and Supplement syllabus content will take **Papers 2, 4 and either 5 or 6**. The maximum grade that can be achieved from these papers is A*.

Papers 1 and 2 are multiple-choice papers. These papers:
- take 45 minutes to complete
- consist of 40 items of the four-choice type
- test assessment objectives AO1 (Knowledge with understanding) and AO2 (Handling information and problem solving)
- are weighted at 30% of the final total mark.

Papers 3 and 4 are written papers. These papers:
- take 1 hour 15 minutes to complete
- consist of short-answer and structured questions
- test assessment objectives AO1 and AO2
- are weighted at 50% of the final total mark.

Candidates will take **either** Paper 5 or Paper 6, in which they will not be required to use knowledge outside the Core syllabus content.

Paper 5 is a practical test. The paper:

- takes 1 hour 15 minutes to complete
- tests skills in assessment objective AO3 (Experimental skills and investigations)
- is weighted at 20% of the final total mark.

Paper 6 is an alternative to the practical paper. The paper:

- takes 1 hour to complete
- tests skills in assessment objective AO3
- is weighted at 20% of the final total mark.

Good examination habits

- Make sure you are fully equipped for your examination; ensure you have pens, pencils, a pencil sharpener, a ruler, a rubber, a calculator, a protractor and compasses.
- Draw all diagrams in pencil.
- Draw graphs in pencil; ensure the drawn line is not thicker than the grid lines on the graph paper. Make sure that any best fit line is drawn in one sweeping movement using a ruler for a straight line and free hand for a curve. When drawing a curve, ensure that your wrist is on the inside of the curve. (If necessary, rotate the question paper.) Plot crosses (×) or encircled dots (☉) rather than dots (·) on your graph.
- Always show fully your working for numerical questions to demonstrate your understanding. Remember to include units for calculated quantities in your answers to numerical questions.
- Ensure that you re-read the questions and not just your answers when you have completed your examination. You may find you have the right answer but for an entirely different question.
- Working through past examination questions is an important part of your preparation.

Contents

Important information for students	vi
Cambridge IGCSE Physics	viii
Good examination habits	ix
Revision checklist	xii

Unit 1 General physics
1.1 Length and time	2
1.2 Motion	6
1.3 Mass and weight	16
1.4 Density	22
1.5 Forces	28
1.6 Momentum	48
1.7 Energy, work and power	53
1.8 Pressure	76

Unit 2 Thermal physics
2.1 Simple kinetic molecular model of matter	83
2.2 Thermal properties and temperature	94
2.3 Thermal processes	118

Unit 3 Properties of waves, including light and sound
3.1 General wave properties	125
3.2 Light	131
3.3 Electromagnetic spectrum	146
3.4 Sound	148

Unit 4 Electricity and magnetism
4.1 Simple phenomena of magnetism	155
4.2 Electrical quantities	161
4.3 Electric circuits	186
4.4 Digital electronics	201
4.5 Dangers of electricity	206
4.6 Electromagnetic effects	209

Unit 5 Atomic physics
5.1 The nuclear atom	227
5.2 Radioactivity	231

Additional support material

1. Physical quantities and units — 242
2. How to use formulae effectively — 244
3. Formulae — 245
4. Working with numbers — 252
5. Graphs — 254
6. Glossary — 258
7. Glossary of examination terminology — 263
8. Index — 264

Revision checklist

		Tick box once revised. (The more times you revise each topic the better.)				
		1	2	3	4	5
General physics	1.1 Length and time					
	1.2 Motion					
	1.3 Mass and weight					
	1.4 Density					
	1.5 Forces					
	1.6 Momentum					
	1.7 Energy, work and power					
	1.8 Pressure					
Thermal physics	2.1 Simple kinetic molecular model of matter					
	2.2 Thermal properties and temperature					
	2.3 Thermal processes					
Properties of waves, including light and sound	3.1 General wave properties					
	3.2 Light					
	3.3 Electromagnetic spectrum					
	3.4 Sound					
Electricity and magnetism	4.1 Simple phenomena of magnetism					
	4.2 Electrical quantities					
	4.3 Electric circuits					
	4.4 Digital electronics					
	4.5 Dangers of electricity					
	4.6 Electromagnetic effects					
Atomic physics	5.1 The nuclear atom					
	5.2 Radioactivity					

"Everything should be made as simple as possible, but not simpler."

Attributed to Albert Einstein 1879 – 1955

Unit 1 General physics

Section 1.1 Length and time

❑ When making measurements, physicists use different instruments, such as rulers to measure length, measuring cylinders to measure volume and stopwatches to measure time.

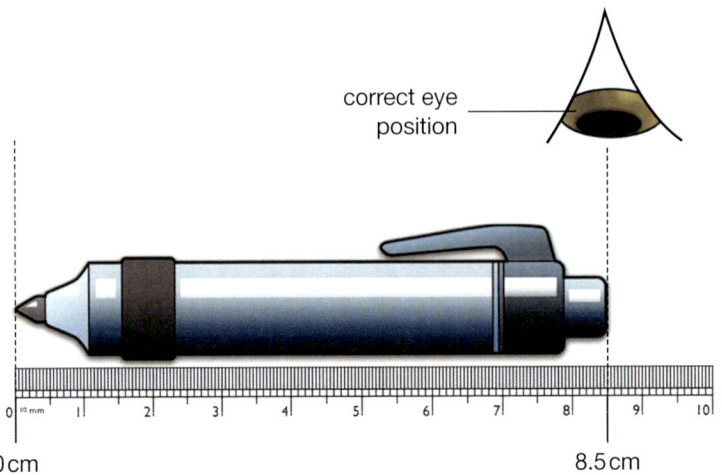

❑ A ruler is used to measure length. When using a ruler, be careful to avoid **parallax** error.

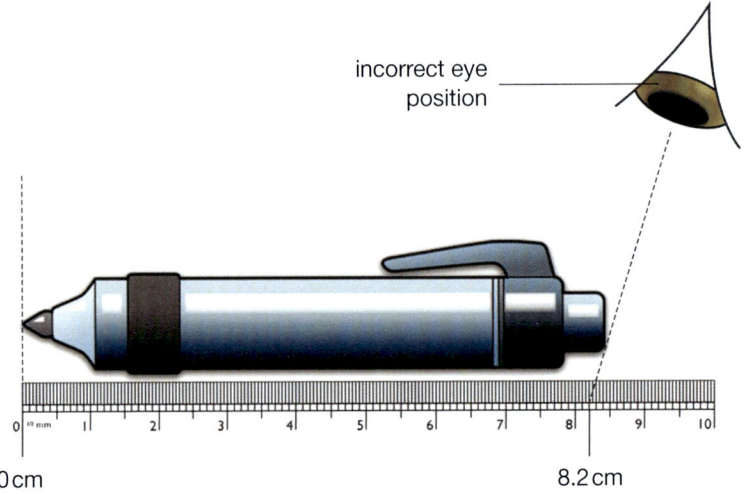

> **Note**
>
>  **Parallax** causes an object to appear shorter or longer depending on how you view it. An object must be viewed at right angles to the scale to measure its length correctly.

❑ A measuring cylinder is used to measure volume. Volume must be measured to the bottom of the **meniscus** (the bottom of the curved surface of liquid).

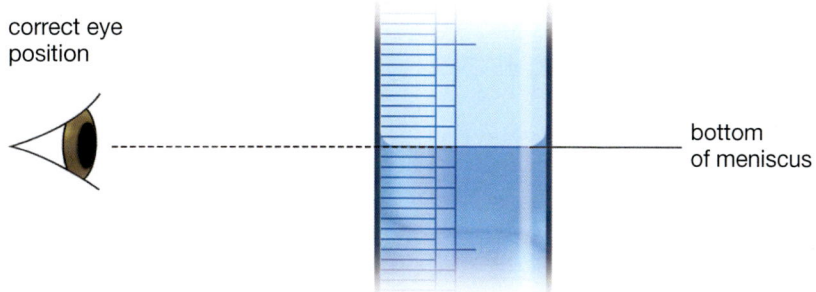

❑ When using a measuring cylinder, again be careful to avoid parallax error.

❑ If the volume is not measured at right angles to the meniscus, parallax error will cause incorrect high or low measurements.

○ A micrometer screw gauge is a mechanical device that is used to measure small lengths very **precisely**.

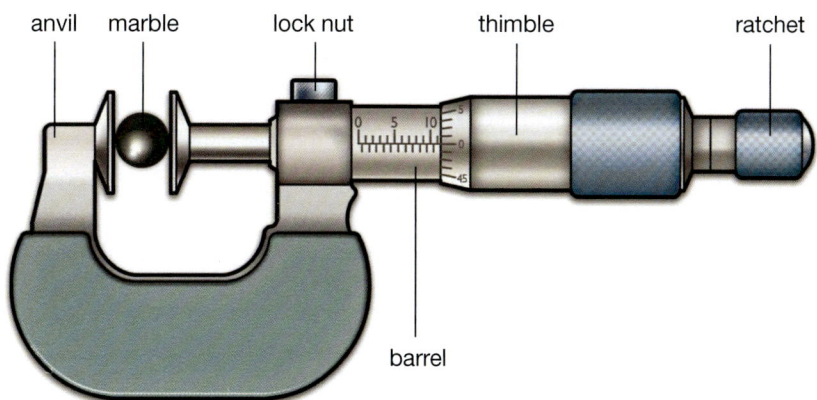

○ To measure the diameter of a marble, open the jaws of the micrometer and then close them gently onto the marble. The scale on the barrel is in mm. The scale on the thimble is calibrated from 0 to 50, where 50 is equivalent to a further 0.50 mm.

○ The diagrams show readings of a micrometer for two different objects. Adding the values on the three scales gives the final reading for the size of the object.

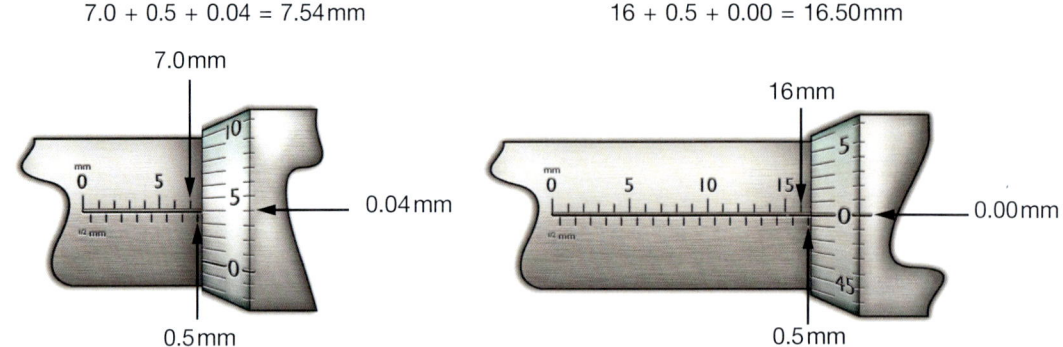

❑ To get a more **accurate** value for such small measurements, repeat the process several times and calculate an average of your readings.

If you have, say 50 sheets of paper, it is possible to measure the thickness of all 50 sheets and then calculate the thickness of one sheet by dividing by 50.

Similarly, if you measure the height of a column of 20 coins you can calculate the thickness of one coin by dividing the height of the column by 20.

❑ Measuring the **period** of one pendulum swing can be very difficult, especially if the arc of the swing is small. The time it would take you to react would affect the measurement.

❑ By measuring the time taken for many (10 or 20) complete swings (oscillations), the period can be calculated by dividing the total time taken by the number of swings.

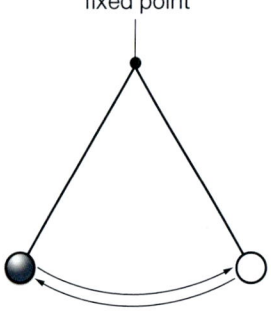

one complete swing
(all the way to the right
and back to the left)

$$period = \frac{total\ time}{number\ of\ swings}$$

Unit 1 General physics

- ❑ Time can be measured with an analogue stopwatch or clock as shown on the previous page. These are mechanical devices as they operate from a coiled spring, which operates a series of gear wheels as it unwinds.

- ❑ A digital stopwatch can measure time to 0.01 of a second and so will give a **more precise** reading than using a mechanical stopwatch with a sweep second hand. It could still be **inaccurate** because of the reaction time of the person using it.

- ❑ The digital stopwatch below shows a time of 3 minutes and 43.00 seconds.

- ❑ Digital timers are often used in connection with electronic circuits and can be switched on and off by sensors. They can therefore measure time intervals with high accuracy as well as high precision.

Top Tip

Repeating measurements, identifying and removing any results that are obviously wrong, and calculating an **average** or **mean value** is likely to improve the **accuracy** of an experiment.

Section 1.2 Motion

❑ **Speed** is a measure of how fast something is moving or the distance travelled per unit of time.

❑ **Average speed** is the speed measured over a relatively long period of time.

$$\text{average speed} = \frac{\text{total distance}}{\text{total time}}$$

❑ **Instantaneous speed** is the speed measured over an extremely short period of time.

❑ **Speed** has the symbol *v* or sometimes *u*, and its unit is the **metre per second** (m/s).

❑ **Distance** has the symbol *d*, and is measured in **metres** (m).

❑ **Time** has the symbol *t*, and is measured in **seconds** (s).

❑ These quantities are related by the formula:

$$d = vt$$

d = distance (m)
v = speed (m/s)
t = time (s)

❑ *Example*
A car travels at 20 m/s for 1 minute and 10 seconds. Calculate how far it travels.

Step 1 List all the information in symbol form and change into appropriate and consistent SI units if required.

v = 20 m/s
t = 1 minute and 10 seconds = 70 s
d = ?

Step 2 Use the correct formula.

$d = vt$

Step 3 Calculate the answer by putting the numbers into the formula.

$d = vt = 20 \times 70 = 1400\,m = 1.4\,km$

ALWAYS REMEMBER TO STATE THE UNIT FOR CALCULATED QUANTITIES.

○ A **scalar** quantity has **magnitude** (size) only.
A **vector** quantity has **magnitude** (size) and **direction**.

○ **Speed** is a scalar quantity because it has magnitude (size) only and can be described as the distance covered per unit time.

○ **Velocity** is a vector quantity because it has magnitude and direction. It can be described as the speed in a particular direction.

○ The table below gives examples of scalars and vectors:

Scalar
speed
time
distance
energy
mass
power

Vector
velocity
acceleration
displacement
force
weight
momentum

Distance–time graphs

- ❏ Stationary
  (Not moving, at rest)

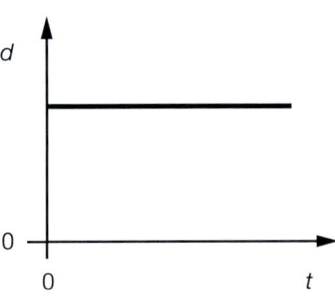

- ❏ Constant speed
  (No acceleration)

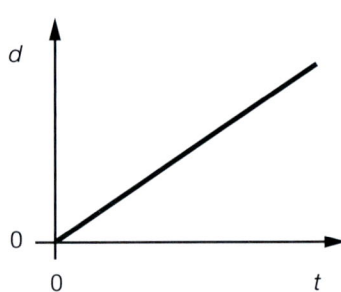

- ❏ Increasing speed
  (Acceleration)

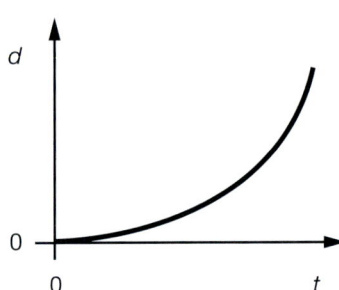

- ❏ Decreasing speed
  (Deceleration)

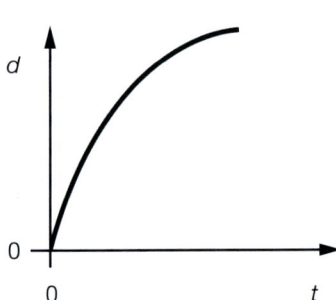

- ○ The speed can be calculated from the **gradient** (slope) of a distance–time graph.

Note

In mathematics the gradient is given by:

$$m = \frac{\Delta y}{\Delta x}$$

m = gradient
Δy = change in y
Δx = change in x

For a distance–time graph:

$$\text{gradient} = \frac{\text{distance travelled}}{\text{time taken}} = \text{speed}$$

Speed–time graphs

- ❏ Constant speed
  Steady (uniform) speed
  No acceleration
  At rest if $v = 0$

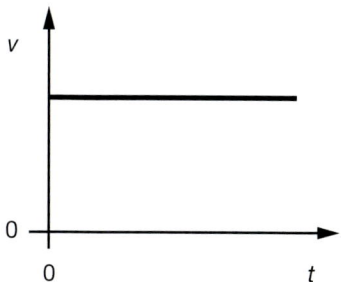

- ❏ Constant acceleration
  Speeding up

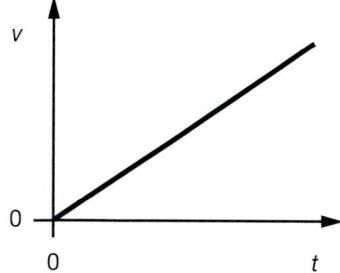

- ❏ Constant deceleration
  Slowing down
  Negative acceleration
  Decreasing speed

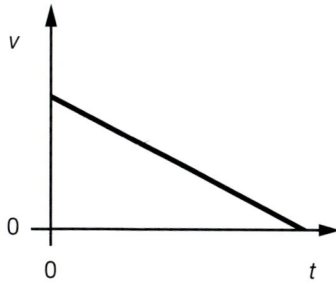

- ○ Changing acceleration
  Increasing acceleration
  (Gradient of curve increases)

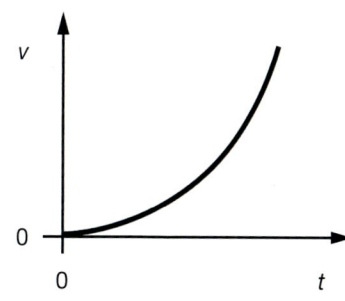

- ○ Changing acceleration
  Decreasing acceleration
  (Gradient of curve decreases)
  Not decreasing speed

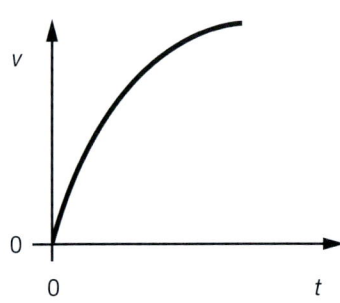

- ○ Changing deceleration
  Decreasing deceleration
  (Gradient of curve decreases)
  Decreasing speed

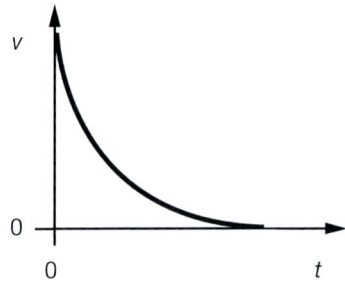

- ○ Acceleration has the symbol **a**, and it is the rate of change of velocity (how quickly an object becomes faster or slower). Its unit is the **metre per second squared** (m/s²).

$$a = \frac{v - u}{t}$$

a = acceleration (m/s²)
v = final speed (m/s)
u = initial speed (m/s)
t = time (s)

The information from a **speed–time graph** can be used to calculate various values.

- ○ The **acceleration** – which is calculated by dividing the change in velocity by the time taken.

- ❑ The **distance travelled** – which is equal to the **area under** the speed–time graph.

- ○ The **maximum acceleration** – which can be found by choosing the part of the graph with the steepest gradient (steepest slope) and then calculating the gradient.

Note

Remember: the gradient is calculated by

$$m = \frac{\Delta y}{\Delta x} \quad \begin{array}{l} m = \text{gradient} \\ \Delta y = \text{change in } y \\ \Delta x = \text{change in } x \end{array}$$

This time the gradient **m** gives the acceleration. A negative gradient on a speed–time graph shows that the object is slowing down or decelerating.

Example

The speed–time graph below represents a motorbike going on a very short journey from the rider's house (A) to a local store (F). Use the graph to:

❑ (a) describe the motion at the various stages of the journey AB, BC, CD, DE and EF

○ (b) calculate the acceleration at the various stages of the journey

❑ (c) state the maximum speed during the journey

❑ (d) calculate the total distance travelled.

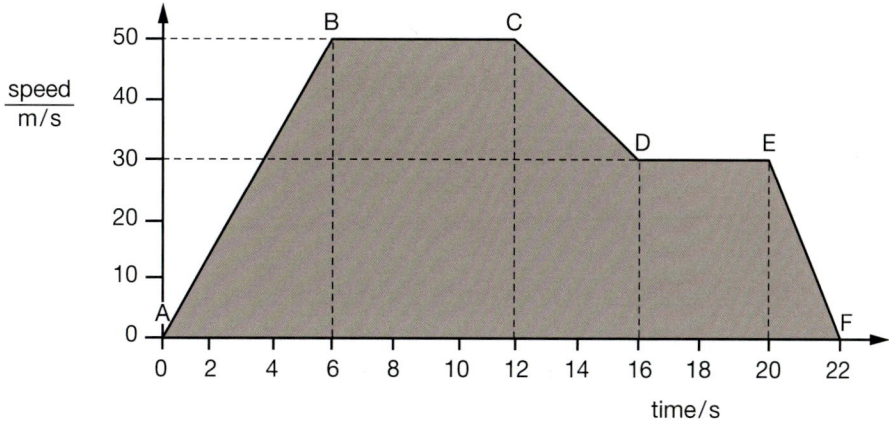

(a) Stage AB: constant acceleration
Stage BC: constant speed
Stage CD: constant deceleration
Stage DE: constant speed
Stage EF: constant deceleration

(b) AB: $a = \dfrac{v - u}{t} = \dfrac{50 - 0}{6.0} = 8.3 \text{m/s}^2$

CD: $a = \dfrac{v - u}{t} = \dfrac{30 - 50}{4.0} = -5.0 \text{m/s}^2$

EF: $a = \dfrac{v - u}{t} = \dfrac{0 - 30}{2.0} = -15 \text{m/s}^2$

For BC and DE $a = 0$

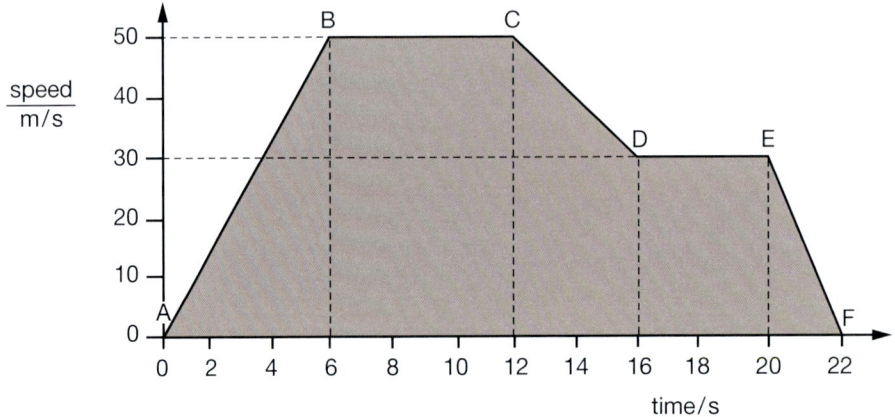

(c) Maximum speed = 50 m/s

(d) Distance travelled = area under the graph

AB: $\dfrac{1}{2}(6.0 \times 50) = 150 \text{m}$

BC: $(6.0 \times 50) = 300 \text{m}$

CD: $\dfrac{1}{2}\left[(50 - 30) \times 4.0\right] + (4.0 \times 30) = 40 + 120 = 160 \text{m}$

DE: $(4.0 \times 30) = 120 \text{m}$

EF: $\dfrac{1}{2}(30 \times 2.0) = 30 \text{m}$

Total distance = 150 + 300 + 160 + 120 + 30 = 760 m

ALWAYS REMEMBER TO STATE THE UNIT FOR CALCULATED QUANTITIES.

○ The acceleration formula can be rearranged as:

$$v = u + at$$

a = acceleration (m/s²)
v = final speed (m/s)
u = initial speed (m/s)
t = time (s)

○ When the **inital speed u** is zero, the formula can be more simply expressed as $v = at$.

○ **Example**
A toy car accelerates from rest at 10 cm/s² for 18 s. Calculate the final speed.

Step 1 List all the information in symbol form and change into appropriate and consistent SI units if required.

$u = 0$ because the toy car **starts from rest**
$a = 10\,\text{cm/s}^2 = 0.10\,\text{m/s}^2$
$t = 18\,\text{s}$
$v = ?$

Step 2 Use the correct formula.

$v = u + at$
$v = at$ because $u = 0$

Step 3 Calculate the answer by putting the numbers into the formula.

$v = at$
$= 0.10 \times 18 = 1.8\,\text{m/s}$

ALWAYS REMEMBER TO STATE THE UNIT FOR CALCULATED QUANTITIES.

In an examination situation you will often be required to think further using pre-existing knowledge.

Example
A quad bike decelerates uniformly from 3.5 m/s to 1.0 m/s in 10 s. Calculate the distance travelled by the bike.

There are two ways to solve this problem.

○   Method 1 – Calculation

   Step 1   List all the information in symbol form and change into appropriate and consistent SI units if required.

   $u = 3.5$ m/s
   $v = 1.0$ m/s
   $t = 10$ s
   $d = ?$

   Step 2   Use the correct formula.
   Because the deceleration is uniform:

   $$\text{distance} = \text{average speed} \times \text{time} = \left(\frac{u+v}{2}\right)t$$

   Step 3   Calculate the answer by putting the numbers into the formula.

   $$d = \left(\frac{u+v}{2}\right)t = \left(\frac{3.5+1.0}{2}\right) \times 10 = 22.5 \text{ m}$$

ALWAYS REMEMBER TO STATE THE UNIT FOR CALCULATED QUANTITIES.

○ <u>Method 2 – Using a graph</u>

Step 1 Sketch a speed–time graph using the information given.

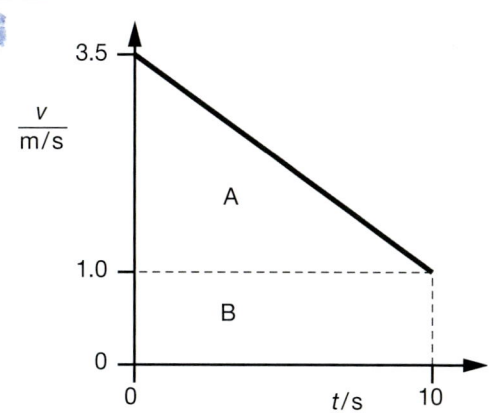

Step 2 Calculate the area under the graph. This is equal to the distance travelled.

Area A = $\frac{1}{2}\left[(3.5-1.0)\times 10\right]$ = 12.5 m

Area B = 10 × 1.0 = 10 m

Area A + Area B = 12.5 + 10 = 22.5 m

ALWAYS REMEMBER TO STATE THE UNIT FOR CALCULATED QUANTITIES.

Section 1.3 Mass and weight

- Weight has the symbol **W** and it is a gravitational force. Its unit is the **newton** (N).

- Force and hence weight can be measured using a newton meter, also known as a spring balance.

- The **weight** of an object is defined as the force due to gravity acting on an object's mass.

- The **mass** of an object is a measure of the amount of matter in the object and can be measured using a mass balance.

- Mass has the symbol **m** and its unit is the **kilogram** (kg).

○ Mass is a property that resists change in motion. The greater the mass the more difficult it is to set an object in motion, to stop it or to change its direction.

○ **Inertia** is defined as the **reluctance** of an object to move or change its motion. Objects with a large mass require larger forces to move them than objects with a small mass. A speedboat requires less force to make it change direction than an oil tanker.

- The mass of an object will not change as the strength of gravity changes. The weight, however, will change as the strength of gravity changes.

- The **gravitational field strength** is the weight per unit mass; it has the symbol **g** and its unit is the **newton per kilogram** (N/kg).

$$g = \frac{W}{m}$$

W = weight (N)
m = mass (kg)
g = gravitational field strength (N/kg)

❑ The relationship between **weight**, **mass** and **gravitational field strength** can be expressed as:

$$W = mg$$

W = weight (N)
m = mass (kg)
g = gravitational field strength (N/kg)

❑ On Earth the gravitational field strength is 10 N/kg.

❑ Because g is a constant near to Earth, the weight of an object is proportional to its mass.

❑ If mass doubles, weight doubles. It follows that weights may be compared using a mass balance.

❑ *Example*
On Earth a man has a mass of 85 kg. Calculate his weight.

Step 1   List all the information in symbol form and change into appropriate and consistent SI units if required.

m = 85 kg
g = 10 N/kg
W = ?

Step 2   Use the correct formula.

$W = mg$

Step 3   Calculate the answer by putting the numbers into the formula.

$W = mg = 85 \times 10 = 850\,\text{N}$

ALWAYS REMEMBER TO STATE THE UNIT FOR CALCULATED QUANTITIES.

> **Top Tip**
>
>
> The **mass** of an object will remain the **same** whether it is on Venus, Mars or Earth or anywhere else. This is because the **amount of matter** in the object stays the **same**. The **weight changes** because the **gravitational field strength** changes on different planets.

- The diagram below shows how the mass remains constant whether an astronaut is on the Moon or on Earth. Only the weight changes due to the different gravitational field strengths.

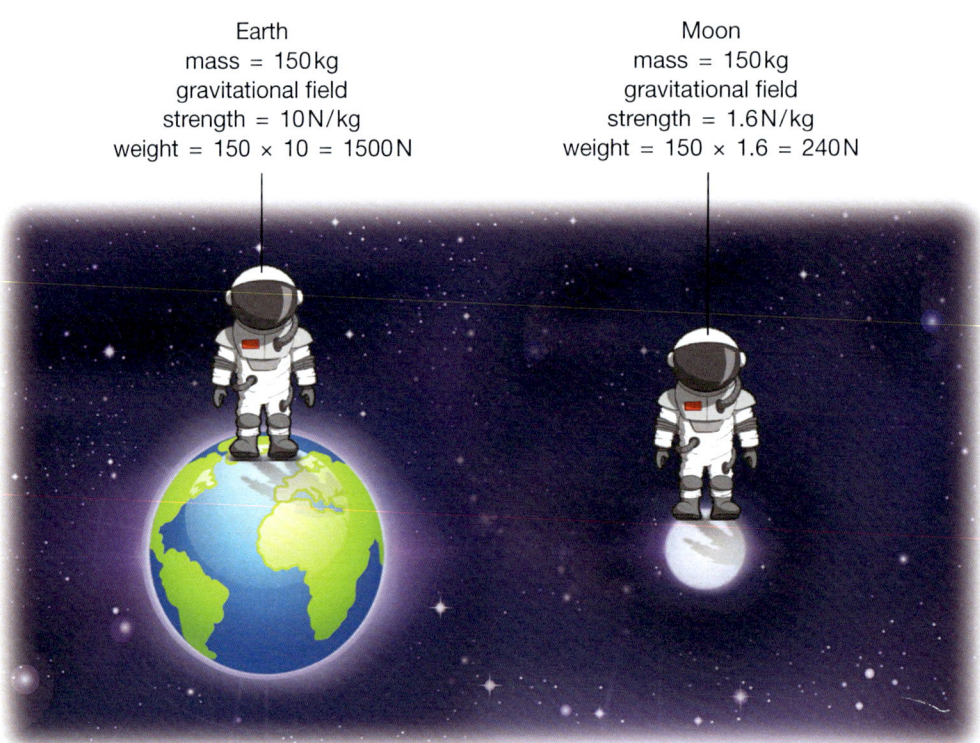

Earth
mass = 150 kg
gravitational field strength = 10 N/kg
weight = 150 × 10 = 1500 N

Moon
mass = 150 kg
gravitational field strength = 1.6 N/kg
weight = 150 × 1.6 = 240 N

Acceleration of free fall

- When an object is dropped from a height and falls to the ground, the force acting on it that causes it to fall is the force due to **gravity**.

- Any **object falling** under gravity will accelerate at 10 m/s². This is known as the **acceleration of free fall**. If there is no **air resistance** then the object will continue to speed up by 10 m/s every second. It will **accelerate**.

- The acceleration of free fall (acceleration due to gravity) for a body falling towards Earth is 10 m/s². This is constant for objects near to the Earth (ignoring any air resistance).

- The acceleration of free fall is numerically equal to the gravitational field strength.

- If there is sufficient air resistance, then the object's acceleration will start at 10 m/s² and then **decrease to zero**, at which time its velocity becomes **constant**. This velocity is called **terminal velocity**.

- Any object **thrown upwards decelerates** at 10 m/s² (ignoring air resistance).

- A student drops two objects of different masses from the leaning tower as shown here. The objects will accelerate at 10 m/s² and hit the ground at the same time provided air resistance is negligible.

- If the student had dropped a feather at the same time as one of the objects shown, it would have been more affected by air resistance and would have quickly reached terminal velocity.

Notes   IGCSE Physics Summarised

- When a parachutist jumps from a small aircraft he begins to fall to the ground as a result of the force due to gravity (his weight). At first his acceleration is 10 m/s².

- The graph opposite shows how his velocity varies with time during his descent.

- As he accelerates, air resistance opposes his motion and decreases his acceleration. At A the gradient of the graph is decreasing. The faster he travels, the greater the air resistance. In other words, the air resistance increases as he falls.

- Eventually he will reach terminal velocity at B when the air resistance is equal to the force due to gravity. Because the two forces are in opposite directions there is no resultant (net) force (see page 28).

- This terminal velocity is too high for him to land safely.

- When he opens his parachute the air resistance increases greatly because of the very large surface area of the parachute.

- The air resistance is greater than the force due to gravity. He continues to fall but decelerates rapidly, as shown at point C.

- As he slows down, the air resistance decreases until once again he reaches terminal velocity at D, when the air resistance equals the force due to gravity. This terminal velocity is much lower and will allow him to land safely.

- When he lands his velocity becomes zero.

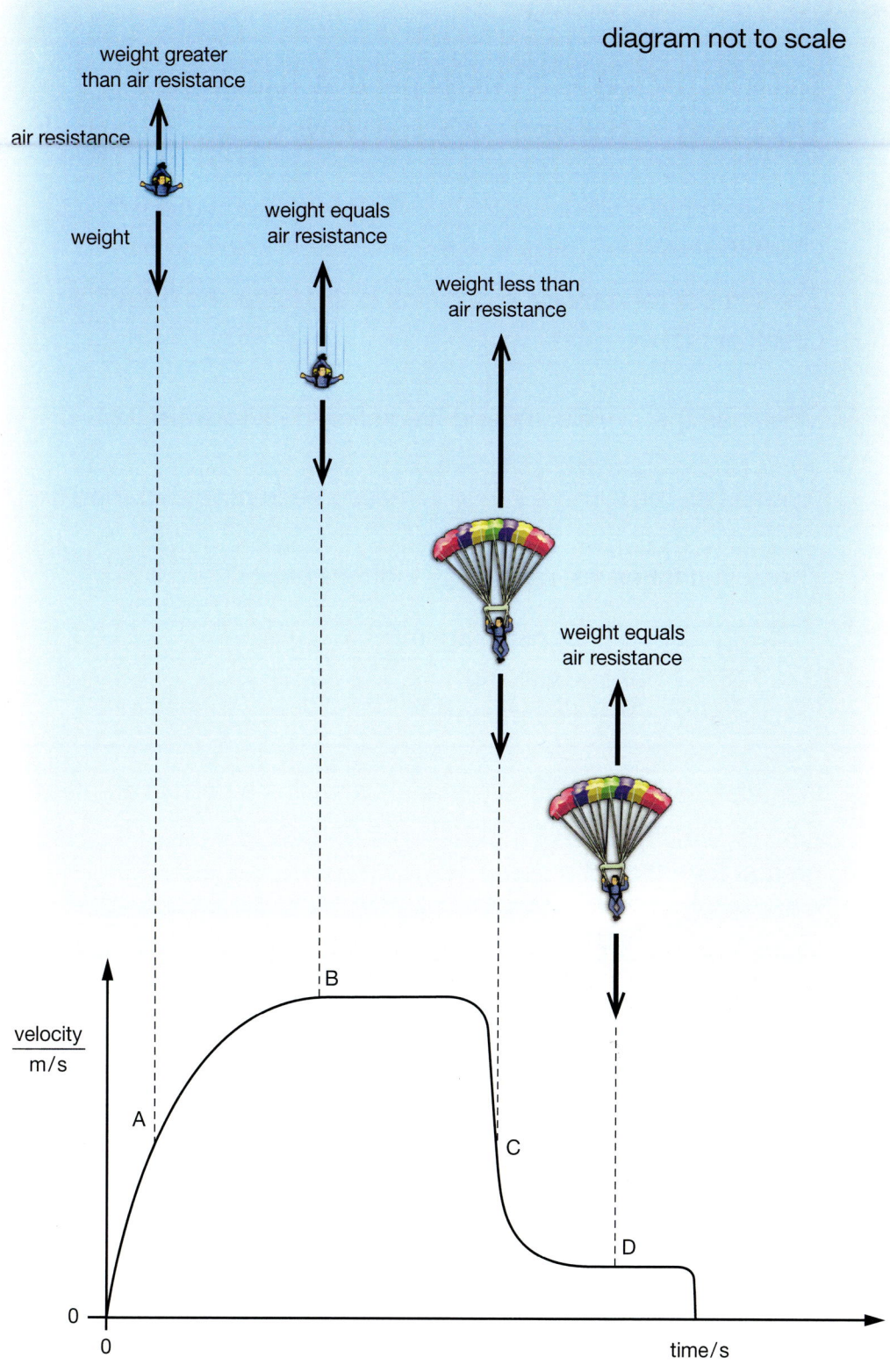

Section 1.4 Density

- Density is defined as the **mass per unit volume** (how much mass is packed into a unit volume).

- Density is measured in kg/m³. (Other common units are g/m³ or g/cm³.)

  The symbol for density is **ρ**, which is the letter rho from the Greek alphabet.

- Mass has the symbol **m**, and its unit is the **kilogram** (kg).

- Volume has the symbol **V**, and is measured in **metres cubed** (m³).

- These quantities are related by the formula:

  $$\rho = \frac{m}{V}$$

  ρ = density (kg/m³)
  m = mass (kg)
  V = volume (m³)

- Generally, the density of a material when it is a solid is greater than its density when it is liquid. The density of a liquid is greater than that of a gas.

- Water is an exception because the density of the solid, ice, is lower than that of water. Ice floats in water because it is less dense.

- Lead is more dense than water and sinks. Some wood is less dense than water and floats.

- A liquid of lower density will float on top of a liquid of higher density, e.g. oil floats on water.

> **Note**
>
>  Each substance has its own density; one bar of gold will have the **same density** as 100 bars of gold.

❑ Consider the following diagram representing gas particles in a box. All the particles have the same mass.

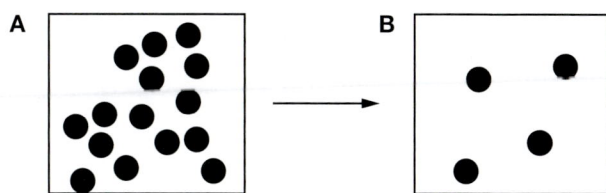

Boxes A and B have the same volume but a different number of particles. Box A has higher density as it has more particles.

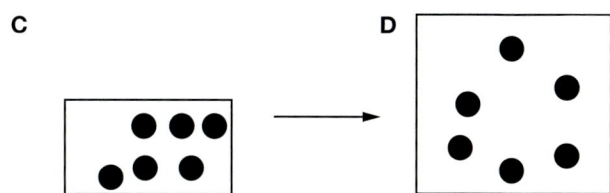

Box C is half the volume of box D with the same number of particles; hence box C has the higher density. The density of a gas depends on the number of particles (the mass) and the volume.

Determining density

❑ **Determining the density of a regularly shaped object**

1. Use a mass balance to measure the mass *m* of the object.

2. Use a ruler to measure the dimensions of the object and then calculate its volume **V**.

3. Then use the following formula to calculate density:

$$\rho = \frac{m}{V}$$

> **Note**
>
>  If you are using a spring balance (newton meter) instead of a mass balance then you are measuring the weight **W**. Calculate the mass from the weight value using the formula:
>
> $$m = \frac{W}{g}$$ where *g* is the gravitational field strength (10 N/kg).

❑ **Example**

The block of wood has a mass of 80g. Calculate its density.

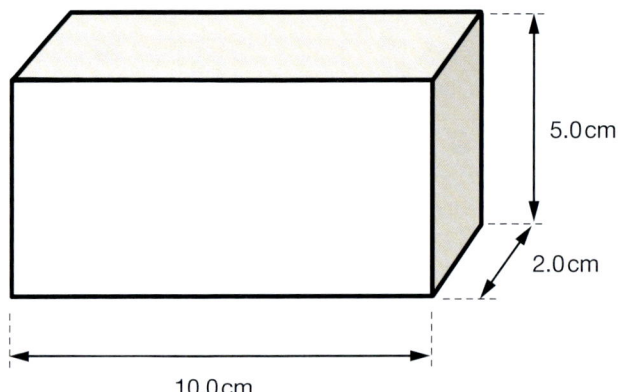

Step 1 List all the information in symbol form and change into appropriate and consistent SI units if required.

$l = 10.0\,cm$
$w = 2.0\,cm$
$h = 5.0\,cm$
$m = 80\,g$
$\rho = ?$

Step 2 Use the correct formulae.

$V = l \times w \times h$ and $\rho = \dfrac{m}{V}$

Step 3 Calculate the answer by putting the numbers into the formulae.

$V = l \times w \times h = 10.0 \times 2.0 \times 5.0 = 100\,cm^3$

$\rho = \dfrac{m}{V} = \dfrac{80}{100} = 0.80\,g/cm^3$

ALWAYS REMEMBER TO STATE THE UNIT FOR CALCULATED QUANTITIES.

Both kg/m^3 and g/cm^3 are acceptable units.

Unit 1 General physics

❏ **Determining the density of a liquid**

1. The mass *m* of the liquid can be measured using a mass balance. To find the mass of the liquid we subtract the mass of the empty measuring cylinder from the mass of the liquid and the measuring cylinder.

2. The volume *V* can be read directly from the measuring cylinder.

3. Then use the following formula to calculate density.

$$\rho = \frac{m}{V}$$

❏ **Example**

Use the information in the diagram below to calculate the density of water.

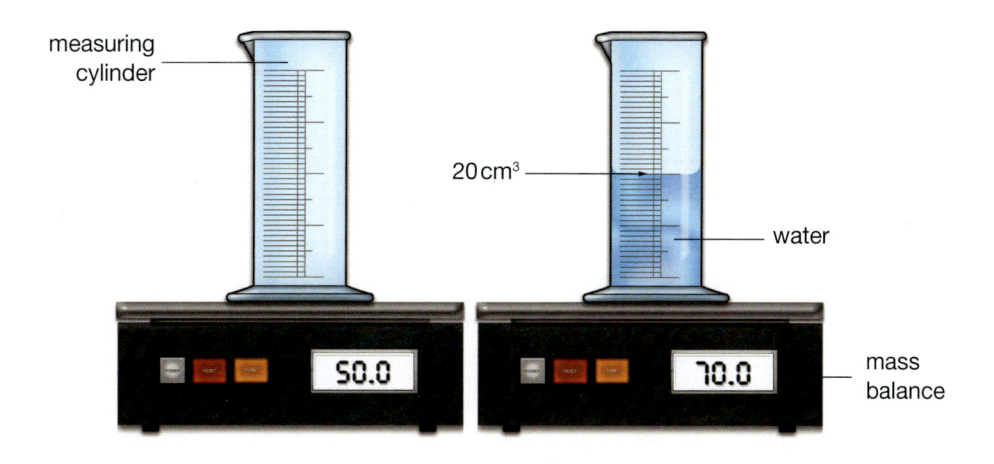

- measuring cylinder
- 20 cm³
- water
- 50.0
- 70.0
- mass balance

Step 1   List all the information in symbol form and change into appropriate and consistent SI units if required.

$m = 70 - 50 = 20 \text{g}$
$V = 20 \text{cm}^3$
$\rho = ?$

It is acceptable to use g and cm³ for density calculations.

Step 2   Use the correct formula.

$$\rho = \frac{m}{V}$$

Step 3 Calculate the answer by putting the numbers into the formula.

$$\rho = \frac{m}{V} = \frac{20}{20} = 1.0 \, g/cm^3$$

ALWAYS REMEMBER TO STATE THE UNIT FOR CALCULATED QUANTITIES.

❏ **Determining the density of an irregularly shaped object**

In the procedure below we use the **displacement** of water method to determine the volume **V**.

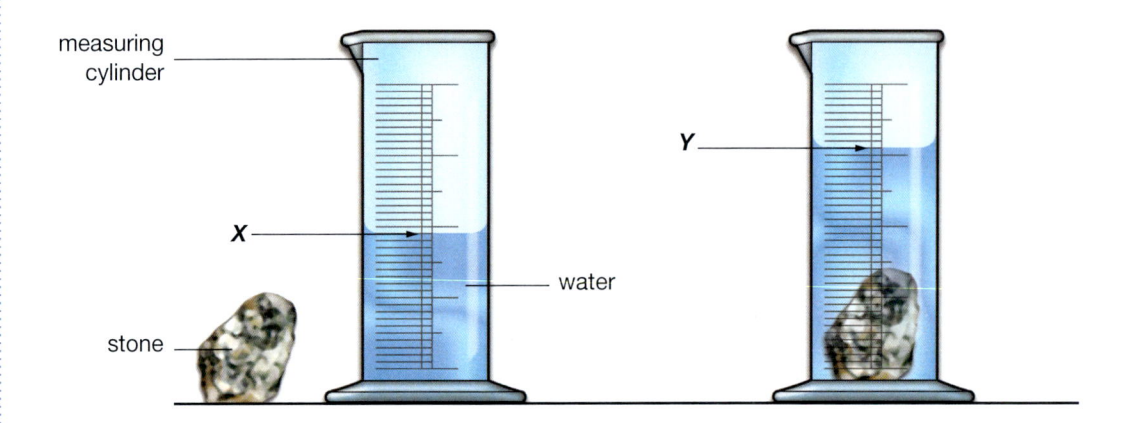

1. Measure the mass **m** of the irregularly shaped object, in this case a stone, using a mass balance.

2. Partially fill a measuring cylinder with a known volume **X** of water.

3. Immerse the stone into the water.

4. Measure the new volume **Y**.

5. The volume of the stone is **Y − X**.

6. Use the following formula to calculate the density

$$\rho = \frac{m}{V}$$

Remember: Always measure the bottom of the **meniscus** when using a measuring cylinder.

Unit 1 General physics   Notes

❑ **Example**

An irregularly shaped piece of steel has a mass of 100g. The steel is immersed in the water as shown. Using the information from the diagram, calculate the density of steel.

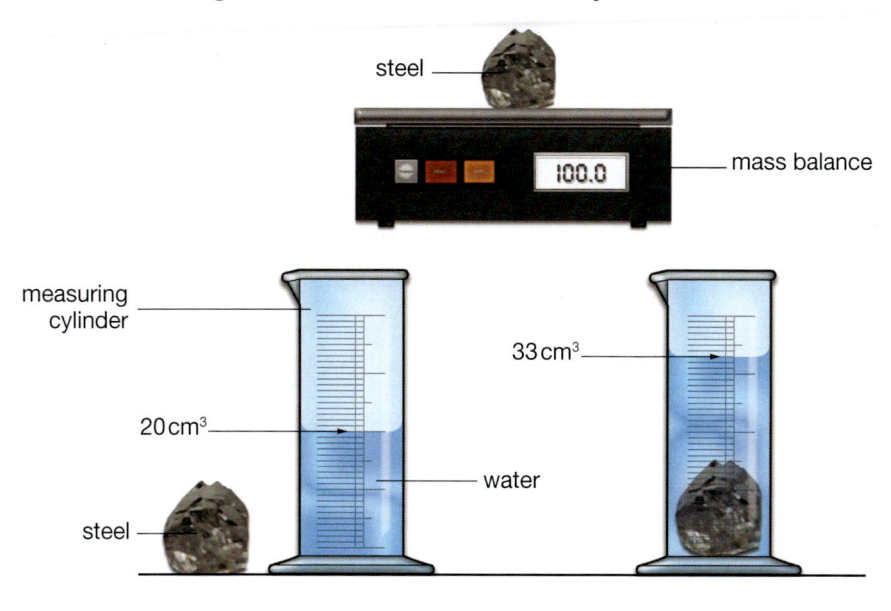

Step 1 List all the information in symbol form and change into appropriate and consistent SI units if required.

$m = 100\,g$
$V = 33 - 20 = 13\,cm^3$
$\rho = ?$

Step 2 Calculate the density by putting the numbers into the formula.

$$\rho = \frac{m}{V} = \frac{100}{13} = 7.7\,g/cm^3$$

ALWAYS REMEMBER TO STATE THE UNIT FOR CALCULATED QUANTITIES.

> *Top Tip*
>
> If an irregularly shaped object floats on water, use a pin or a sharp pencil to poke it down below the surface so you can find its volume. Be careful not to push the pin or pencil into the water otherwise you will be measuring its volume as well as that of the irregular object.

Section 1.5 Forces

- ❏ A **force** is a pull, push, twist, stretch, squeeze, tug or shove. We can observe the effects of forces but cannot see the actual forces themselves.

- ❏ Forces can:
  - change the speed of an object
  - change the direction of movement of an object
  - change the shape of an object
  - change the size of an object.

- ○ Force has magnitude and direction, so it is a vector.

- ❏ If the forces acting in opposite directions are equal, the forces are said to be **balanced**. The sum of the forces is zero in such cases. (In the following diagram all the values of **F** are the same.)

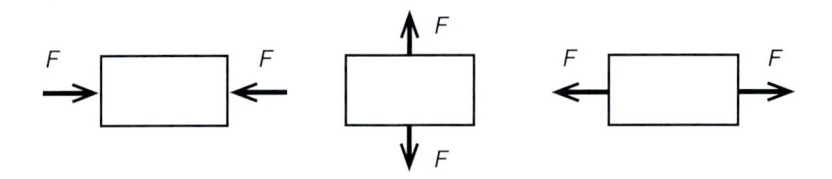

$$F \rightarrow \boxed{} \leftarrow F \qquad \begin{array}{c} \uparrow F \\ \boxed{} \\ \downarrow F \end{array} \qquad F \leftarrow \boxed{} \rightarrow F$$

- ❏ When there are **no forces** acting, or the forces acting on an object are **balanced**, the object may be **stationary**.

- ❏ When an object is moving at a **constant** speed in a straight line, there is no **resultant force**. The forces on it are equal and in opposite directions.

- ❏ If the forces on an object are not balanced there will be a **resultant force** that will cause the object to speed up, slow down or change direction, depending on the direction of the resultant force. The resultant force is the **overall unbalanced force**.

- ○ **Resultant force** is the vector sum of all the forces on an object.

This leads to the following laws:

- **Newton's 1st Law** states that an object remains **at rest** or moves at a **steady speed** in a **straight line** unless acted on by a **resultant** or **unbalanced force**.

- **Newton's 2nd Law** states that an object **accelerates** in the direction of a **resultant** or **unbalanced force**.

  $$F = ma$$

  F = resultant force (N)
  m = mass (kg)
  a = acceleration (m/s²)

- Resultant force has the symbol F and its unit is the **newton** (N).

- Mass has the symbol m, and is a measure of the amount of matter in an object. Its unit is the **kilogram** (kg).

- Acceleration has the symbol a, and it is the rate of change of velocity (how quickly an object becomes faster or slower). Its unit is the **metre per second squared** (m/s²).

- When the resultant force applied to an object is constant and the **mass** of the object **increases**, the **acceleration decreases**.

- When the mass of an object is constant and the **resultant force** applied to the object **increases**, the **acceleration increases**.

- The two formulae for acceleration in the Cambridge IGCSE Physics syllabus are:

  $$a = \frac{F}{m} \qquad a = \frac{v - u}{t}$$

Circular motion

○ If an object is moving in a **circle**, or along the **arc** of a circle, there must be a **force** acting on it to continually change its **direction**.

○ The force, which always acts **towards the centre** of the circle, is given the name **centripetal force**. It acts **perpendicularly** (at right angles) to the direction of motion of the object at any given instant.

○ In circular motion, a force always acts towards the centre. This ensures that the object **accelerates** towards the centre but **does not move** towards the centre. Objects A and B are examples of objects experiencing centripetal force.

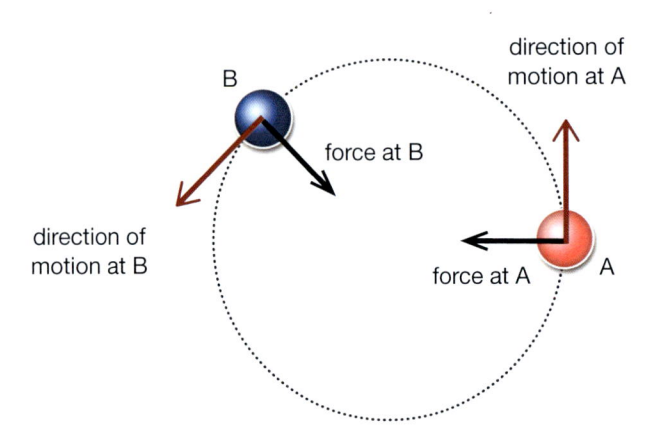

○ Satellites orbiting the Earth, the Earth orbiting the Sun, a tennis ball attached to a string and moving in a circle, are all examples where **centripetal forces** are present.

Top Tip

During circular motion an object's direction is always changing, so its **velocity** is always changing even when its **speed** remains the same (constant).

Unit 1 General physics

Adding and subtracting to calculate resultant forces

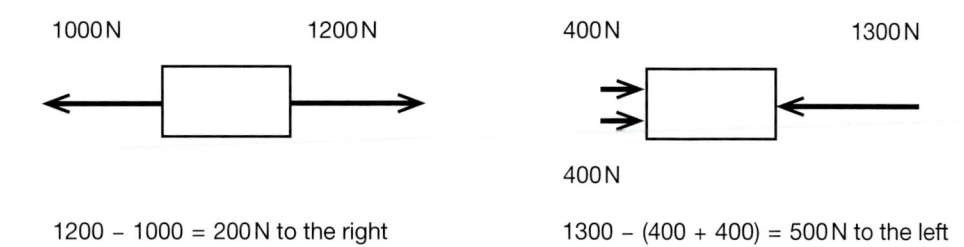

1000 N ← □ → 1200 N

1200 − 1000 = 200 N to the right

400 N ⇒ □ ← 1300 N
400 N ⇒

1300 − (400 + 400) = 500 N to the left

- Force is a vector, so it always has a direction.

- If forces are acting in opposite directions they should be subtracted. If forces are acting in the same direction they should be added.

- The same principles are used for forces acting vertically upwards and downwards, e.g. the forces on a falling object.

- **Example**

  A jet ski accelerates at 2.0 m/s² and has a resultant force of 540 N in the direction of motion. Calculate the mass of the jet ski. (The mass of the rider may be ignored.)

  Step 1 List all the information in symbol form and change into appropriate and consistent SI units if required.

  $F = 540 \text{ N}$
  $a = 2.0 \text{ m/s}^2$
  $m = ?$

  Step 2 Use and rearrange the correct formula.

  $F = ma \quad \Rightarrow \quad m = \dfrac{F}{a}$

  Step 3 Calculate the answer by putting the numbers into the formula.

  $m = \dfrac{F}{a} = \dfrac{540}{2.0} = 270 \text{ kg}$

  ALWAYS REMEMBER TO STATE THE UNIT FOR CALCULATED QUANTITIES.

Friction

- ❑ The force of friction **opposes the motion** of an object. Air resistance (drag) is a form of friction.

- ❑ Friction causes heating, e.g. when one material is pushed across the surface of another.

- ❑ Friction can be useful in certain situations and not very useful in others.

- ❑ Friction is **useful** in the following situations and can be **increased** by:
    - a skydiver opening a parachute allowing him/her to slow down quickly due to increased air resistance
    - pressing the brake pedals in a car, slowing it down quickly.

- ❑ Friction is **not useful** in the following situations and can be **decreased** by:
    - skiers putting wax on their skis to make them smooth
    - making objects **streamlined**, allowing them to travel faster by cutting through the air
    - oiling engines to allow the parts to move easily.

Unit 1 General physics   Notes

○ **Example**

A Formula One car has a mass of 640 kg (including the driver). When the engine exerts a driving force of 29 000 N, the opposing frictional force is 800 N. Calculate the acceleration.

Step 1 Draw a diagram showing the forces acting on the object and their directions.

800 N ← [640 kg Formula One car] → 29 000 N

Step 2 Calculate the **resultant (unbalanced)** force.

[640 kg Formula One car] → 28 200 N

$F = 29000 - 800 = 28200$ N

Step 3 List all the information in symbol form and change into appropriate and consistent SI units if required.

$F = 28200$ N
$m = 640$ kg
$a = ?$

Step 4 Use and rearrange the correct formula.

$$F = ma \implies a = \frac{F}{m}$$

Step 5 Calculate the answer by putting the numbers into the formula.

$$a = \frac{F}{m} = \frac{28\,200}{640} = 44.06 \text{ m/s}^2 = 44 \text{ m/s}^2$$

ALWAYS REMEMBER TO STATE THE UNIT FOR CALCULATED QUANTITIES.

Hooke's Law

○ Hooke's Law states that the **extension x** of a spring is **proportional** to the **force F** applied. This means that if the force is doubled, the extension is doubled.

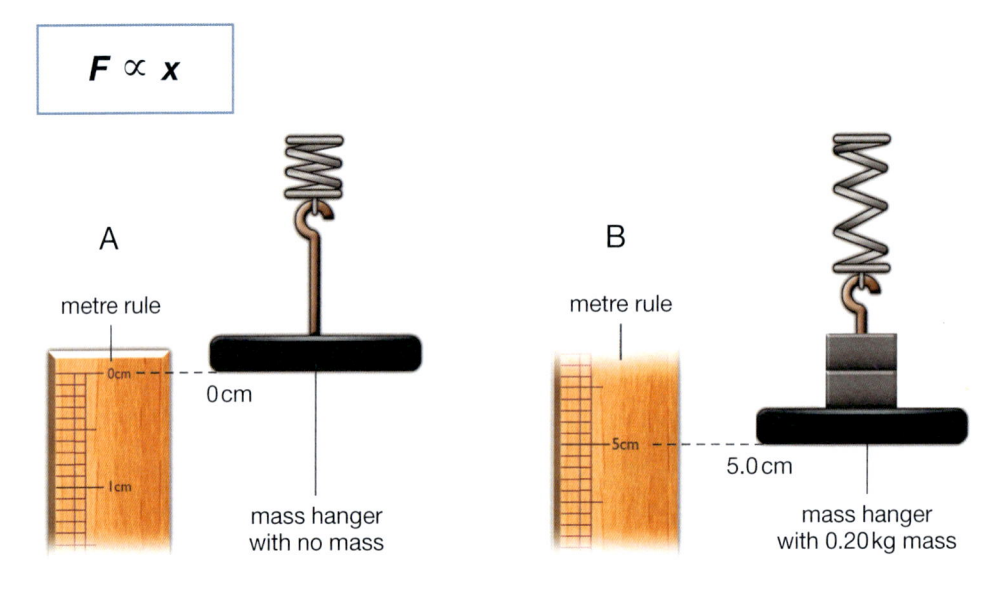

$$F \propto x$$

A
metre rule
0 cm
0 cm
1 cm
mass hanger with no mass

B
metre rule
5 cm
5.0 cm
mass hanger with 0.20 kg mass

❏ The following experiment investigates Hooke's Law using a spring.

1. Assemble the apparatus as shown in diagram A. (In a laboratory environment a clamp stand, metre rule, spring, mass hanger and slotted masses would be needed.)

2. Note and record the reading of the scale of the rule next to the bottom of the mass hanger without adding any masses. In diagram A this is 0 cm.

3. Add one slotted mass (the weight or load of a 100 g mass is 1.0 N) to the hanger and measure the extension on the ruler. The extension is measured by

   extension = new length − original length

4. Repeat step 3, adding one mass at a time and record the corresponding extension reading.

(Diagram B shows two 100g masses added. In this case the extension would equal the length of the spring with the 2.0N load minus the length of the spring with no load which is 5.0 − 0 = 5.0 cm.)

5. Prepare a table of your results for load (calculated from the mass readings) and extension.

6. Plot a graph of extension against load. Your graph should be a straight line passing through the origin. This shows that the extension is proportional to the load (weight or force).

○ The spring should return to its original length if you remove the masses.

○ If you continued to add masses your graph would look like the one below. The **limit of proportionality** is shown and is where the straight line ends.

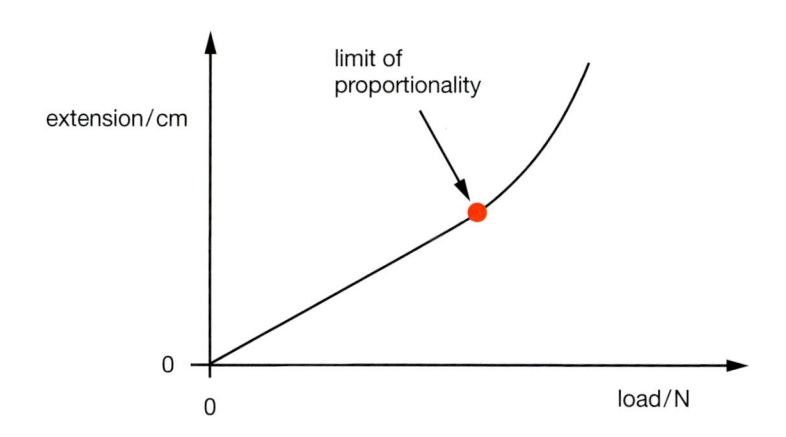

> ### Note
> If any two quantities are **directly proportional**, when they are plotted against each other on a graph, the graph has two characteristics.
> - It passes through the **origin**.
> - It is a **straight line**.
>
> For Hooke's Law, if the load (force) is doubled then the extension doubles up to the limit of proportionality. In other words, the gradient of the extension–load graph is the same up until this point.

IGCSE Physics Summarised

- Force has the symbol **F** and its unit is the **newton** (N).

- Extension of a spring has the symbol **x** and is measured in mm, cm or m.

- The stiffness of the spring or the **spring constant** has the symbol **k** and is measured in N/mm, N/cm or N/m.

- These quantities are related by the formula:

    $$F = kx$$

    F = force (N)
    k = spring constant (N/m)
    x = extension (m)

- The **spring constant k** can be calculated by rearranging the above formula:

    $$k = \frac{F}{x}$$

    The **limit of proportionality** is the point beyond which the spring extension **will not be proportional** to the load. Up to this limit the extension increases by a set amount for every newton of force applied. Above this limit the increase in extension per newton will be greater.

- The limit of proportionality is an important point.
    - If the spring is stretched beyond this point, it will no longer extend proportionally to the load (weight) applied.
    - If the spring is stretched only up to this point, the extension will be proportional to the load (weight) applied.

- **Example**
  A spring is stretched by 0.030 m by a load of 3.0 N. Assuming the limit of proportionality is not reached, calculate:
  (a)   the spring constant of the spring
  (b)   the load required to stretch the spring by 0.080 m.

(a) **Step 1** List all the information in symbol form and change into appropriate and consistent SI units if required.

$F = 3.0\,N$
$x = 0.030\,m$
$k = ?$

Step 2 Use and rearrange the correct formula.

$$F = kx \quad \Rightarrow \quad k = \frac{F}{x}$$

Step 3 Calculate the answer by putting the numbers into the formula.

$$k = \frac{F}{x} = \frac{3.0}{0.030} = 100\,N/m$$

ALWAYS REMEMBER TO STATE THE UNIT FOR CALCULATED QUANTITIES.

(b) **Step 1** List all the information in symbol form and change into appropriate and consistent SI units if required.

$k = 100\,N/m$ (answer to part (a))
$x = 0.080\,m$
$F = ?$

Step 2 Use the correct formula.

$$F = kx$$

Step 3 Calculate the answer by putting the numbers into the formula.

$$F = kx = 100 \times 0.080 = 8.0\,N$$

ALWAYS REMEMBER TO STATE THE UNIT FOR CALCULATED QUANTITIES.

Turning effect

- A **moment** is a measure of the **turning effect** of a force.

- The turning effect depends on two things:
    - the **magnitude** of the force applied
    - the **perpendicular distance** of the force from the **pivot**.

- Moment may be given the symbol **M**, and its unit is the **newton metre** (Nm).

- Force has the symbol **F**, and its unit is the **newton** (N).

- Distance has the symbol **d**, and its unit is the **metre** (m).

- These quantities are related by the formula:

$$M = Fd$$

M = moment (Nm)
F = force (N)
d = perpendicular distance (m)

- The moment is greater if either **F** or **d** is increased. For example, you can use a longer spanner to make it easier to undo a tight nut, or you can apply more force on the original spanner.

- Examples where moments are important include:
    - door handles
    - see-saws (or teeter-totters)
    - cranes
    - levers.

> **Note**
>
> Doors are difficult to open or close if you pull or push near the pivot (the hinges); the further from the pivot you try to open close the door, the easier it is. **The longer the distance from the pivot, the smaller the force required.**

❑ A pivot is an object around which rotational movement takes place.

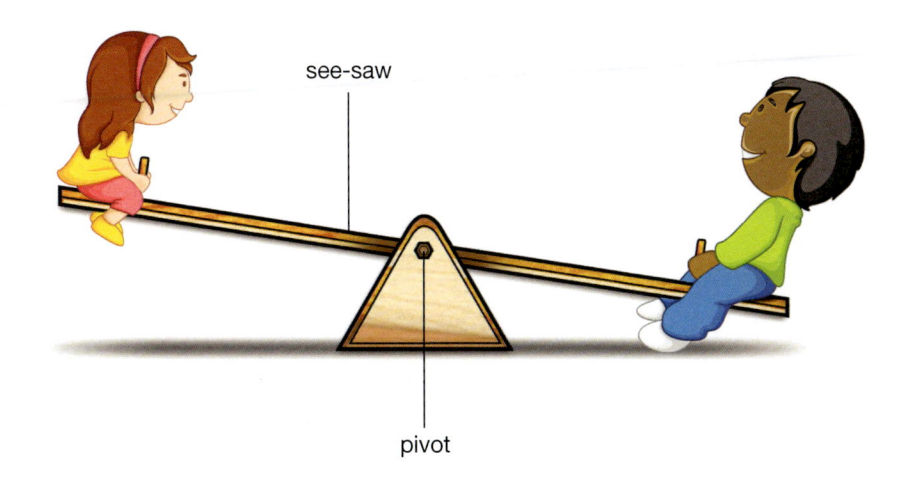

❑ The **principle of moments** states that for a beam to be balanced about a pivot, the sum of the clockwise moments **equals** the sum of the anti-clockwise moments.

> sum of clockwise moments = sum of anti-clockwise moments
> $$F_2 d_2 = F_1 d_1$$

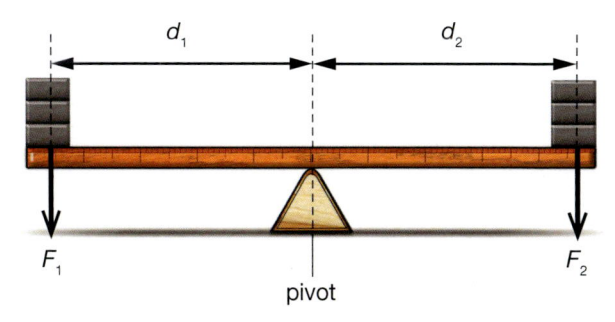

F_1 causes an anti-clockwise turning effect.
F_2 causes a clockwise turning effect

❑ When there is no resultant force and no resultant moment (turning effect), a system will be in **equilibrium (balanced)**.

❏ The following experiment demonstrates the principle of moments.

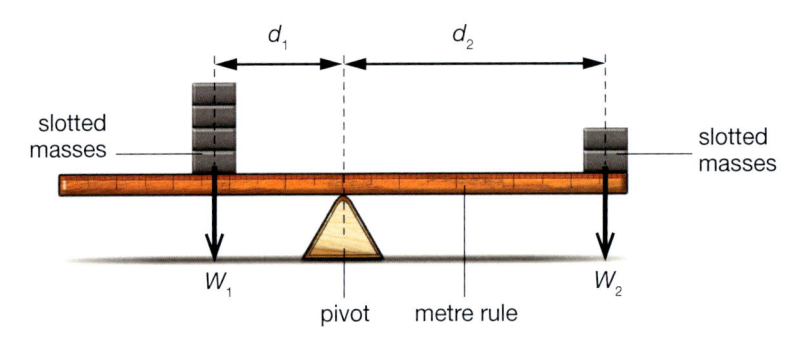

A typical experimental arrangement is shown above.

Apparatus:
- one triangular prism to act as the pivot
- one metre rule
- slotted masses, 100g each (of weight 1.0N each).

1. Balance the metre rule on the pivot at the 50.0cm mark of the rule.

2. Add different masses (of weights) W_1 and W_2 at different distances d_1 and d_2 from the pivot. Carefully adjust the distances d_1 and d_2 until the metre rule balances horizontally.

3. Record the values of W_1, W_2, d_1 and d_2.

4. Repeat steps 2 and 3 several times, with different values for W_1, W_2, d_1 and d_2.

5. For each set of results, calculate the turning moments $W_1 \times d_1$ and $W_2 \times d_2$. You will see that $W_1 d_1 = W_2 d_2$. The metre rule is balanced when the clockwise moment equals the anti-clockwise moment, thus demonstrating the principle of moments. There is no net turning effect when a body is in equilibrium.

Remember: that weight is a force, but we use W instead of F in this case.

The values below are a sample set of results showing that the anti-clockwise moment = the clockwise moment.

W_1/N	d_1/m	W_2/N	d_2/m	$W_1 \times d_1$/Nm	$W_2 \times d_2$/Nm
1.0	0.50	2.0	0.25	0.50	0.50
2.0	0.30	3.0	0.20	0.60	0.60

Note

If there is no net moment and no net force, a system is said to be in equlibrium.

○ **Example**

The diagram below shows an experimental set-up for investigating the moment (turning effect) of a force. The metre rule shown is balanced and therefore in equilibrium.

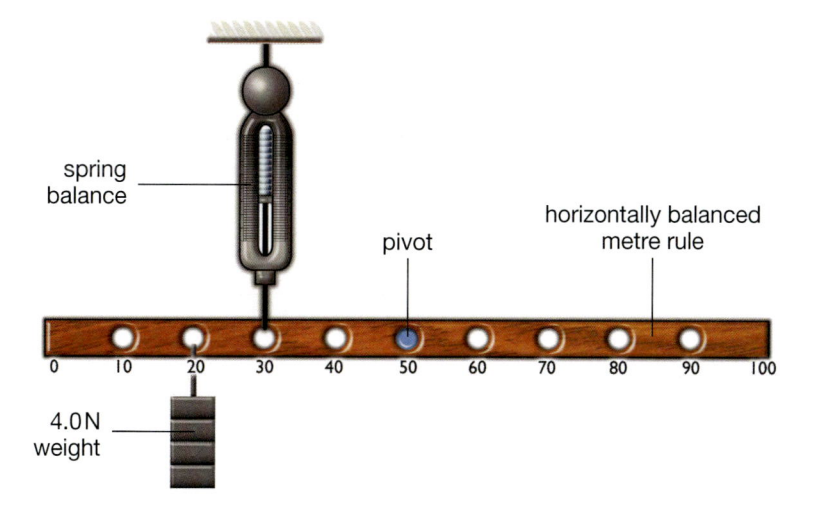

(a) State two conditions for the metre rule to be in equilibrium.
(b) Show that the reading on the spring balance is 6.0 N.
(c) The weight of the rule is 1.7 N. Calculate the force exerted by the pivot on the metre rule.

(a) 1. sum of clockwise moments
= sum of anti-clockwise moments
i.e. there is no net moment on a body in equilibrium.

2. upward force = downward force
i.e. there is no net force on a body in equilibrium.

(b) W_1 is 4.0 N and it is 30 cm (0.30 m) from the pivot.
W_2 is the reading on the spring balance, which is 20 cm (0.20 m) from the pivot.

$$W_1 d_1 = W_2 d_2$$
$$4.0 \times 0.30 = W_2 \times 0.20$$
$$W_2 = \frac{4.0 \times 0.30}{0.20} = 6.0 \, \text{N}$$

(c) upward force = downward force

The upward force is 6.0 N (reading on the spring balance).

The downward force is
4.0 N + 1.7 N (weight + weight of rule) = 5.7 N

Therefore a 0.30 N downwards force is required to balance the other forces, and so this is the force exerted by the pivot on the metre rule.

ALWAYS REMEMBER TO STATE THE UNIT FOR CALCULATED QUANTITIES.

Centre of mass

❑ The centre of mass of an object is the point through which all the weight appears to be acting.

❑ This is a useful simplification as we can assume that the force of **gravity** only acts on a **single point**.

❑ This means a single arrow can represent the weight **W** of an object, as shown in the diagram of the tower below.

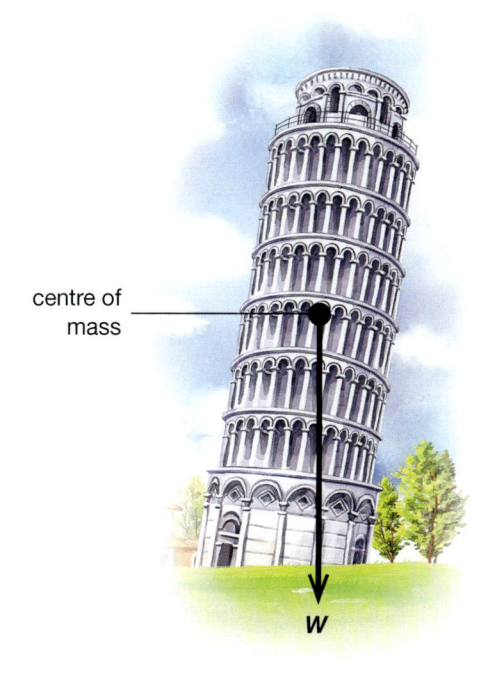

centre of mass

W

❑ Although the tower is leaning, it does not topple because the line of action of its weight falls within the base (see page 45).

❑ The centre of mass for objects with a regular shape is in the centre. Drawing dashed lines from various points through the middle of the shape can easily help find the centre of mass as shown below.

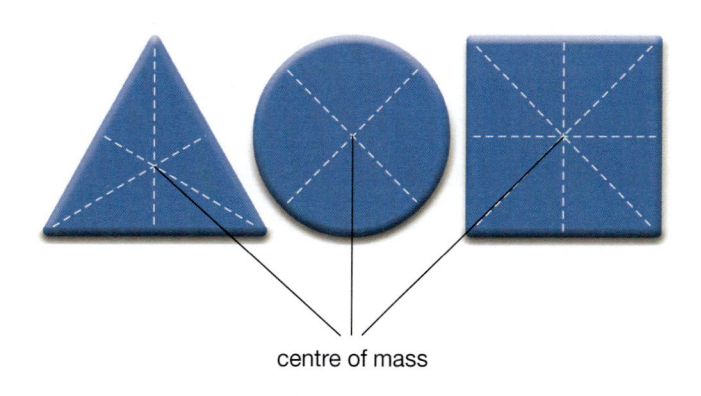

centre of mass

IGCSE Physics Summarised

❏ The following experiment describes how to find the centre of mass of an **irregular** plane shape such as a piece of card.

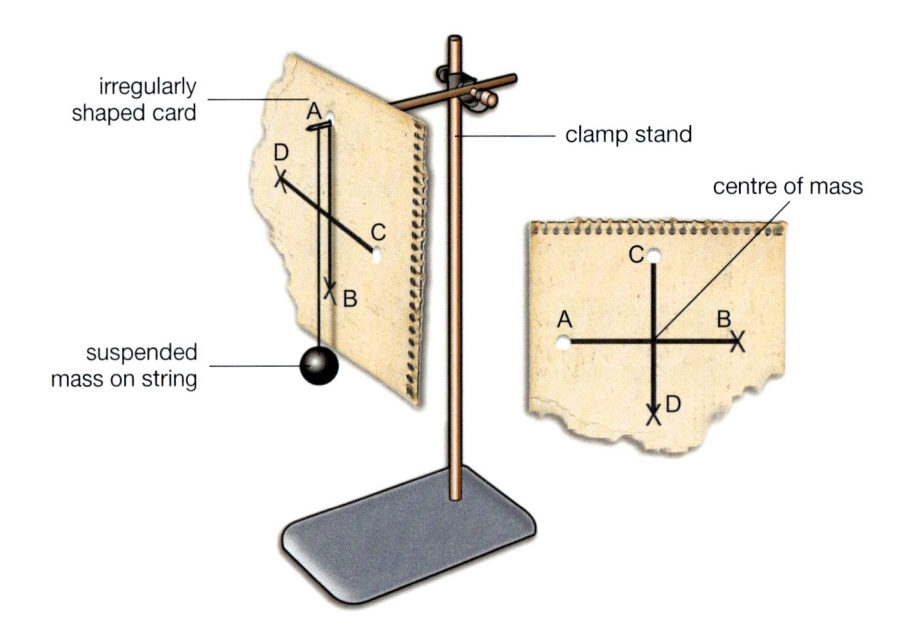

1. Hang the card from a clamp stand from hole A.

2. Suspend a mass on a string from the same place.

3. Put a cross towards the bottom of the card directly behind the string (e.g at point B).

4. Remove the card and draw a line to mark the position of the string (e.g. from A to B). The centre of mass is somewhere along the line of the string.

5. Repeat steps 1 to 4 with the card hanging from different positions (e.g. from hole C). The centre of mass is where the two lines cross.

> **Top Tip**
>
> The centre of mass is the point through which all the weight appears to act. A ruler will balance perfectly when its centre is placed directly over a pivot. Otherwise it will tip over.

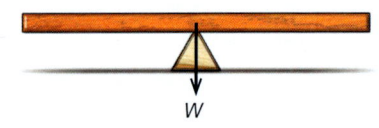

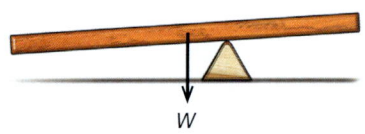

❑ The idea of centre of mass is useful when predicting whether or not an object will topple (fall) over.

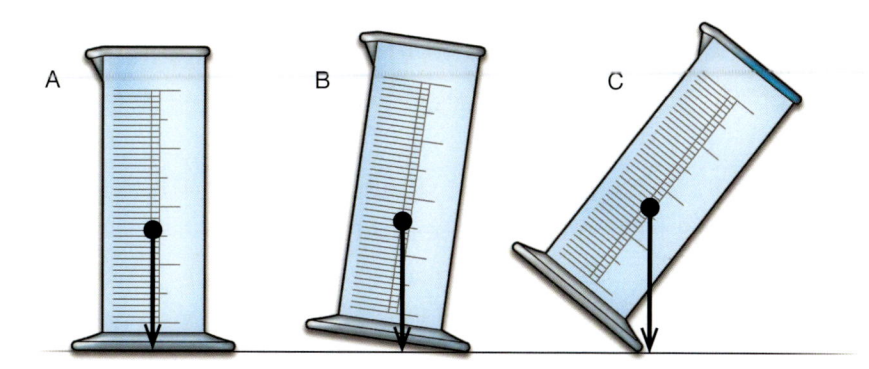

Consider measuring cylinders A, B and C.
- Measuring cylinders A and B will not topple over as the line of action of their weight falls within the base.
- Cylinder A is said to be in **stable equilibrium** as there is no resultant moment.
- Although cylinder B will not topple over, it is not in equilibrium. If it is released it will move back to the upright position.
- Cylinder C will topple over as its line of action of its weight falls outside the base.

❑ A balanced object that is easy to topple over is said to be in **unstable equilibrium**, e.g. an inverted cone balanced on its point. When it is balanced the line of action of its weight acts through the point of the cone. A slight movement will cause the line of action of its weight to fall outside the point and the cone will topple.

> **Note**
>
> Racing cars travel at very high speeds and need to be very stable. They have a small height, which gives them **a low centre of mass**. A racing car is also wide so that if it starts to tip, the line of action of its weight stays within the base and it rights itself.
>
> In general:
> **wide and short = stable (low centre of mass)**
> **tall and thin = less stable (high centre of mass)**

Scalars and vectors

The resultant of two vectors at an angle

○   Force, velocity and acceleration are **vector** quantities. Remember a vector has **magnitude (size)** and **direction**.

○   Speed and mass are **scalar** quantities. Remember a scalar quantity has **magnitude (size)** only.

○   We have learned how to add and subtract forces acting in the same or opposite direction (page 31). The **parallelogram rule** is used when determining the resultant of vector quantities (e.g. force or velocity) that act at an angle to one another.

Consider a girl who sets out to cross a river in a canoe travelling due east at 4.0 m/s. There is a current in the river of 3.0 m/s due north. To find out what will happen, we replace the two velocities with a single resultant velocity **R** that will show the direction in which the girl actually travels and her speed.

To find **R** we draw a scale diagram with the velocities in the correct direction, using a suitable scale, such as 1.0 cm : 1.0 m/s.

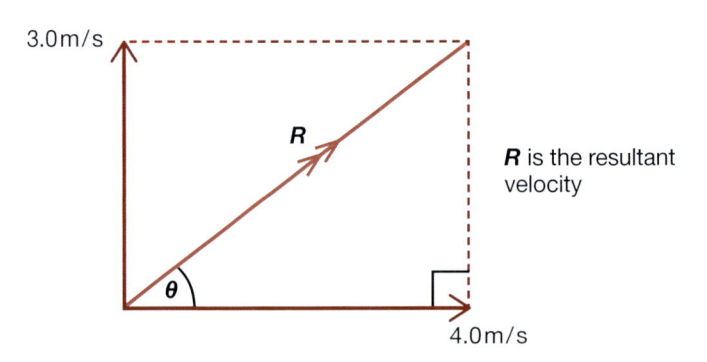

diagram for illustration purposes only, not to scale

○   The velocity of the girl is due east and the velocity of the river is due north. The angle between them is 90°.

The resultant velocity is obtained by completing the parallelogram with dashed lines. An arrow that represents the resultant velocity **R** is then drawn in. It is the diagonal line across the parallelogram from where the velocities originated. Measure the length of this arrow using a ruler and find the size of **R** using the scale of the diagram. You should find it is 5 cm, representing a velocity of 5 m/s.

Measure the angle, θ. You should find it is 37°, indicating that the canoe travels at 37° north of east.

Example

Two tug boats T_1 and T_2 are pulling a larger boat towards the port. T_1 exerts a force F_1 of 3000 N and is travelling due north. T_2 exerts a force F_2 of 5000 N and is travelling at 60° west of north. Use a scale diagram to determine the resultant force and direction of the larger boat.

Step 1 Choose a suitable scale for your diagram, e.g. 1.0 cm : 500 N and convert your force values.

$$3000\,N \Rightarrow 6.0\,cm$$
$$5000\,N \Rightarrow 10.0\,cm$$

Step 2 Using a pencil, ruler and protractor, draw a scale diagram as accurately as you can. Measure the length of the diagonal and convert it back to a force using the scale. Measure the angle θ.

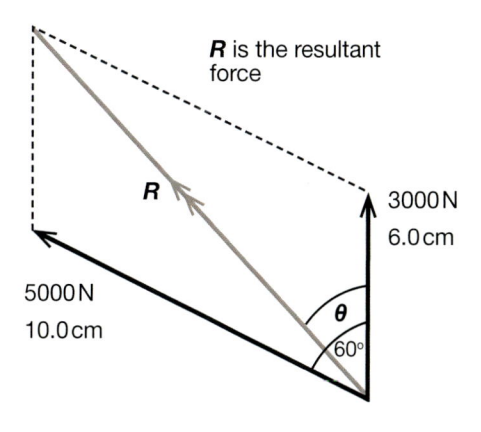

diagram for illustration purposes only, not to scale

You should express your answer as: The resultant force R is 7000 N and the direction of the boat is 38° west of north.

ALWAYS REMEMBER TO STATE THE UNIT FOR CALCULATED QUANTITIES.

Section 1.6 Momentum

- **Linear momentum** is defined as the product of the mass and velocity.

- **Momentum** is a vector quantity; it has size and direction. The momentum of an object will change if either the mass or the velocity changes, and this includes changing direction.

- Mass has the symbol *m*, and its unit is the **kilogram** (kg).

- Velocity has the symbol *v*, and its unit is the **metre per second** (m/s).

- Momentum has the symbol *p*, and its unit is the **kilogram metre per second** (kg m/s).

- These quantities are related by the formula:

$$p = mv$$

p = momentum (kg m/s)
m = mass (kg)
v = velocity (m/s)

- Any mass that is moving has momentum.
    - An oil tanker (large boat) has a large momentum even when it is moving slowly because it has a very large mass.
    - A speedboat has a small momentum in comparison to the oil tanker because, although it can move much faster, it has a much smaller mass.

- Momentum is the tendency of an object to keep moving in the same direction.
    - The same force acting for the same time produces a greater effect on the direction of motion of the speedboat.

○ ***Example***

Calculate the momentum of a car with a mass of 1000 kg travelling due east at 25 m/s. Assume east is the positive direction.

Step 1 List all the information in symbol form and change into appropriate and consistent SI units if required.

$m = 1000\,\text{kg}$
$v = 25\,\text{m/s}$
$p = ?$

Step 2 Use the correct formula.

$p = mv$

Step 3 Calculate the answer by putting the numbers into the formula.

$p = mv = 1000 \times 25 = 25\,000\,\text{kg m/s}$

Momentum is a vector quantity and so the answer should read: The momentum is 25 000 kg m/s due east.

ALWAYS REMEMBER TO STATE THE UNIT FOR CALCULATED QUANTITIES.

○ If we had a similar car travelling at the same speed due west we would say its momentum was −25 000 kg m/s because it is travelling in the opposite direction.

Newton's 2nd Law and momentum

○ We have previously learned (page 29) that according to Newton's 2nd Law:

$$F = ma$$

F = force (N)
m = mass (kg)
a = acceleration (m/s²)

○ The acceleration can be calculated by using:

$$a = \frac{v - u}{t}$$

a = acceleration (m/s²)
v = final speed (m/s)
u = initial speed (m/s)
t = time (s)

○ The acceleration formula can be substituted into $F = ma$ to give:

$$F = \frac{m(v - u)}{t} = \frac{mv - mu}{t}$$

○ Since generally $p = mv$ then, in this case, mv is the final momentum and mu is the initial momentum. Therefore

$$\text{force} = \frac{\text{change in momentum}}{\text{time}}$$

○ Newton's 2nd Law of motion can therefore be worded as: **The rate of change of momentum of an object is equal to the resultant force acting on it.**
This is a different way of saying $F = ma$.

Example

A model train of mass 4.0 kg travels along a straight track. Its velocity increases from 2.0 m/s to 9.0 m/s in 5.0 s. Calculate the average force acting on the train.

Step 1 List all the information in symbol form and change into appropriate and consistent SI units if required.

$m = 4.0 \text{ kg}$
$u = 2.0 \text{ m/s}$
$v = 9.0 \text{ m/s}$
$t = 5.0 \text{ s}$
$F = ?$

Step 2 Use the correct formula.

$$F = \frac{mv - mu}{t}$$

Step 3 Calculate the answer by putting the numbers into the formula.

$$F = \frac{(4.0 \times 9.0) - (4.0 \times 2.0)}{5.0} = 5.6 \text{ N}$$

ALWAYS REMEMBER TO STATE THE UNIT FOR CALCULATED QUANTITIES.

Impulse

The formula below is another useful way of expressing the change in momentum formula where **Ft** is called **impulse**.

Ft = change in momentum = mv − mu

The unit of impulse is the **newton second** (Ns).

> **Note**
>
> It is common practice to denote an object moving from left to right as having a positive velocity and momentum, and an object moving from right to left as having negative velocity and momentum.

Conservation of momentum

○ The **principle of conservation of momentum** states that the total momentum of objects before they collide is equal to their total momentum after the collision provided no external forces act on the objects.

momentum before collision = momentum after collision

This principle allows us to predict how an object will move after a collision. A golf club striking a ball, a meteor crashing into a planet and two cars crashing into one another are all examples of collisions.

A lorry of mass m_1 travelling at a velocity of u_1 collides with a car of mass m_2 travelling at a velocity of u_2. After the collision, their velocities are v_1 and v_2, respectively, as shown below.

$$m_1 u_1 + m_2 u_2 = m_1 v_1 + m_2 v_2$$

Before collision

After collision

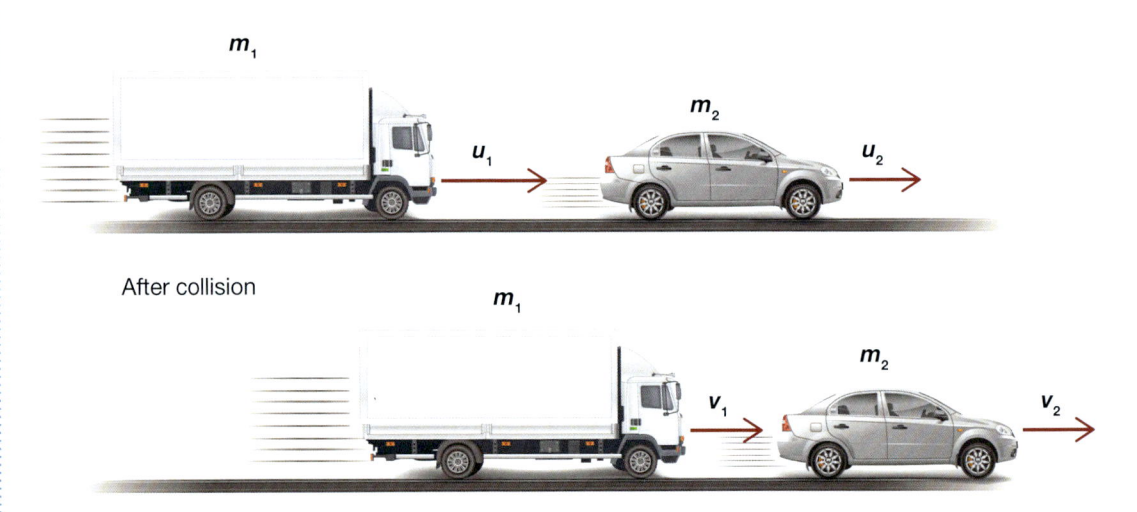

See pages 62–65 for examples involving the conservation of momentum.

Section 1.7 Energy, work and power

Energy

- Energy **cannot** be **created** or **destroyed**; it may change from one form to another. We say that energy is **conserved**. In other words:

    total energy into system = total energy out of system

- The unit of all forms of energy is the **joule** (J).

- There are various forms of energy.

    Kinetic: movement energy
    Potential: stored energy
    Elastic (or strain): stored (i.e. potential) energy when an elastic material is stretched or squashed
    Gravitational potential: energy stored due to height
    Internal (thermal): vibration/movement energy of atoms and molecules, i.e. heat
    Radiation: energy carried as electromagnetic waves, e.g. light
    Electrical: energy transferred by charges, e.g. current
    Chemical: energy stored in fuels such as petrol and food
    Sound: energy transferred as pressure waves through materials
    Nuclear: energy stored inside the nucleus of an atom

- Some examples of **conservation of energy transfers** are:

    1. A bus accelerating along a horizontal road
       chemical energy ➡ kinetic energy + heat + sound
    2. A go-cart braking after travelling at a constant speed
       kinetic energy ➡ heat
    3. A bicycle freewheeling downhill and speeding up
       gravitational potential energy ➡ kinetic energy + heat
    4. A catapult slinging a stone
       elastic (potential) energy ➡ kinetic energy + heat + sound
    5. An electric lamp
       electrical energy ➡ radiation energy (including light) + heat

- ❑ In all changes, some energy is always transferred as **heat** (often due to **friction**) and is often wasted.

Kinetic energy

- ❑ Kinetic energy is the energy associated with **movement** and is abbreviated as **k.e.** Its unit is the **joule** (J).

- ❑ Mass has the symbol **m** and its unit is the **kilogram** (kg).

- ❑ Both **speed** and **velocity** have the symbol **v** and their unit is the **metre per second** (m/s). When speed or velocity increases, **k.e.** increases.

- ○ These quantities are related by the formula

$$\text{k.e.} = \frac{1}{2}mv^2$$

k.e. = kinetic energy (J)
m = mass (kg)
v = velocity (m/s)

To calculate the speed when you know the kinetic energy and mass, rearrange the above formula to give:

$$v = \sqrt{\frac{2(\text{k.e.})}{m}}$$

Remember: The **size** of the velocity **v** is equal to the speed of the object.

> **Top Tip**
>
> When a car travels at a **constant speed**, its **kinetic energy remains constant** as its speed **v** is not changing. Although its **k.e.** is not changing, the car still needs fuel (chemical energy) to keep moving at a constant speed because it is doing work against frictional forces.

Example

A toy car has a mass of 5.0 kg and travels at 60 cm/s. Calculate the kinetic energy of the car.

Step 1 List all the information in symbol form and change into appropriate and consistent SI units if required.

$$v = 60\,\text{cm/s} = 0.60\,\text{m/s}$$
$$m = 5.0\,\text{kg}$$
$$\text{k.e.} = ?$$

Step 2 Use the correct formula.

$$\text{k.e.} = \frac{1}{2}mv^2$$

Step 3 Calculate the answer by putting the numbers into the formula.

$$\text{k.e.} = \frac{1}{2}mv^2 = \frac{1}{2} \times 5.0 \times 0.60^2 = 0.90\,\text{J}$$

(**Remember**: It is only speed that is squared.)

ALWAYS REMEMBER TO STATE THE UNIT FOR CALCULATED QUANTITIES.

Gravitational potential energy

❏ When an object is **lifted**, the work done against gravity is transformed into gravitational potential energy. It is abbreviated as **g.p.e.** or just **p.e.** and its unit is the **joule** (J).

❍ The change in gravitational potential energy is equal to the work done by or against gravity.

❍ The change in gravitational potential energy **g.p.e.** can be expressed by the following formula.

$$\text{g.p.e.} = mg\Delta h$$

g.p.e. = gravitational potential energy (J)
m = mass (kg)
g = gravitational field strength (N/kg)
Δh = change in height (m)

❑ Energy is conserved. If we know the energy of an object at any one point, then calculating the energy at any other point is possible ignoring energy transferred as heat.

○ **Example**
A boy has a mass of 75 kg and dives from a diving board 8.0 m above the surface of the water.

(a) Calculate his gravitational potential energy at A (above that at the water's surface, B) just before he dives.
(b) Find his kinetic energy at B just before he hits the water.
(c) Determine his speed at B.
(d) What energy changes occur after he enters the water?

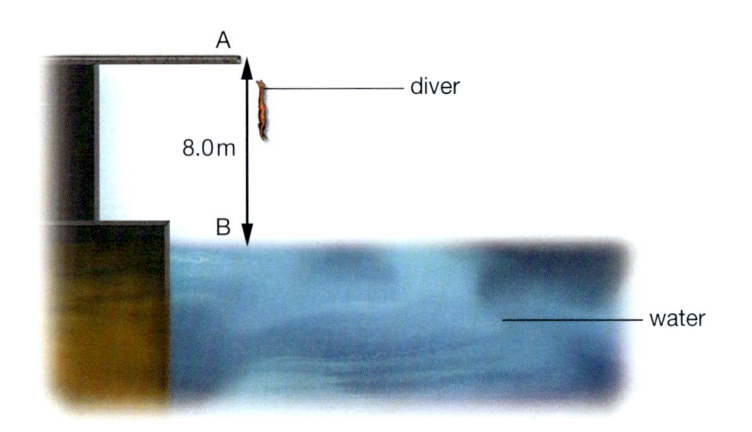

(a) **Step 1** List all the information in symbol form and change into appropriate and consistent SI units if required.

m = 75 kg
Δh = 8.0 m
g = 10 N/kg

Step 2 Use the correct formula.

g.p.e. = $mg\Delta h$

Step 3 Calculate the answer by putting the numbers into the formula.

g.p.e. = 75 × 10 × 8.0 = 6000 J

(b) As he falls, his **g.p.e.** is converted to **k.e.** since energy is conserved.

k.e. = **g.p.e.** = 6000 J

(c) **Step 2** Use and rearrange the correct formula.

$$\text{k.e.} = \frac{1}{2}mv^2 \implies v = \sqrt{\frac{2(\text{k.e.})}{m}}$$

Step 3 Calculate the answer by putting the numbers into the formula.

$$v = \sqrt{\frac{2(\text{k.e.})}{m}} = \sqrt{\frac{2 \times 6000}{75}} = 12.6 \, \text{m/s}$$

ALWAYS REMEMBER TO STATE THE UNIT FOR CALCULATED QUANTITIES.

(d) When the boy enters the water he slows down so he loses kinetic energy, but he causes the water to move so it gains kinetic energy. Some water will splash out and gain gravitational potential energy. You will hear the splash, so some of the energy converts to sound energy and some is wasted when it turns to heat energy in the water.

> **Top Tip**
>
> When an object falls freely in a vacuum under gravity its gravitational potential energy decreases (since it is getting lower) and its kinetic energy increases (since it is getting faster). The **total energy** (**g.p.e.** + **k.e.**) stays the **same** since energy is conserved.

> **Top Tip**
>
> If a runner runs up a hill at constant speed, her **k.e.** remains the same because the speed is **constant**, and her **g.p.e.** increases because she is getting **higher**. Her stored **chemical** energy decreases as she runs and some of the energy is transformed into heat. The **total energy** will remain the **same** since energy is not lost: it just changes from one form into another.

○ When an object falls, the formulae for the change in **g.p.e.** and the final **k.e.** are related as shown below:

g.p.e. = k.e.

$$mg\Delta h = \frac{1}{2}mv^2$$ where Δh is the change in height

The mass *m* can be cancelled from both sides:

$$g\Delta h = \frac{1}{2}v^2$$

This can be rewritten as: $v = \sqrt{2g\Delta h}$

○ The speed or the change in height can be calculated provided we know one of them, as we know that the **gravitational field strength** on Earth is 10N/kg. Consequently we can calculate the final speed or the initial height without knowing the mass.

> **Top Tip**
>
> You could be asked to find the speed of an object dropped from a height. The key here is to understand that the **g.p.e.** at the **top** is equal to the **k.e.** at the **bottom** just before it hits the ground.
>
> **At the top**
> The object has no **k.e.** since it is initially at rest (stationary). It has maximum **g.p.e.** since it is high up.
>
> **Near the bottom**
> When the object is at the bottom, it has maximum **k.e.** since it gets faster as it falls. All its **g.p.e.** has been transferred to **k.e.**

Example

A boy drops a golf ball from rest off a cliff, which is 90m high.

(a) Calculate how fast the ball is travelling just before it hits the ground.

(b) The speed is actually lower in reality than that calculated. Explain why the actual speed is lower than the value calculated in (a).

(a) **Step 1** List all the information in symbol form and change into appropriate and consistent SI units if required.

$h = 90$m
$g = 10$N/kg $= 10$m/s^2
$v = ?$

Step 2 Use the correct formula.

$$v = \sqrt{2g\Delta h}$$

Step 3 Calculate the answer by putting the numbers into the formula.

$$v = \sqrt{2g\Delta h} = \sqrt{2 \times 10 \times 90} = 42.4 \text{m/s}$$

ALWAYS REMEMBER TO STATE THE UNIT FOR CALCULATED QUANTITIES.

(b) The calculation ignores air resistance and assumes that **no energy** is transferred to the surroundings and that **all the gravitational potential energy is transferred into kinetic energy**. In fact, some of the energy will be wasted as heat energy.

- Some energy is **always** transferred as heat; therefore the speed will always be lower than the one calculated, as in the example on the previous page.

- In any process the energy tends to **dissipate** (spread) into the surroundings. When a pendulum bob is pulled to one side it is given gravitational potential energy because it is given height. When it is released this begins to convert to kinetic energy, and then, as it swings past the centre of oscillation, it begins to convert back to potential energy. As time progresses the swings become smaller; some of the potential/kinetic energy is being wasted as heat energy through friction partly due to air resistance. This energy is not recoverable.

- Energy changes are often multi-stage. Consider a hydroelectric power station (see the diagram at the bottom of page 69).
  - The water is stored in a dam at a higher level than the turbines of a generator. It has gravitational potential energy.
  - The water travels through tunnels or pipes to the turbines and its gravitational potential energy changes to kinetic energy as it loses height.
  - As the water causes the turbines to move the kinetic energy of the water converts to kinetic energy of the turbines.
  - The movement of the turbines is transferred to the generator and electricity is produced. The kinetic energy of the turbines is converted to electrical energy.
  - The electrical energy is transferred via power lines to the consumer. It can then be used in many ways.

Remember that the total energy remains the same although some energy will be dissipated or 'lost' to the surroundings in each transfer.

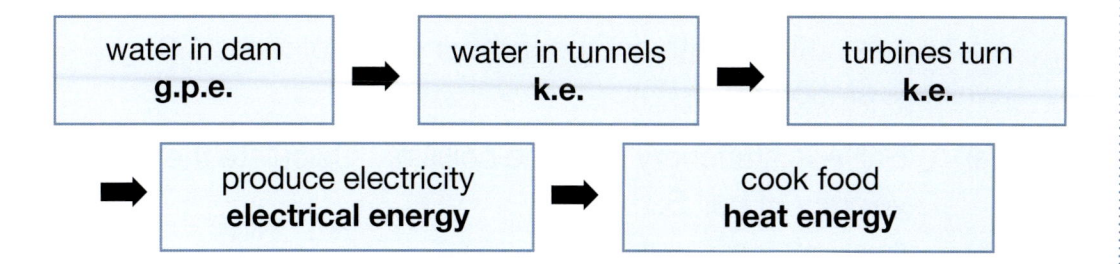

water in dam **g.p.e.** ➡ water in tunnels **k.e.** ➡ turbines turn **k.e.**

➡ produce electricity **electrical energy** ➡ cook food **heat energy**

Momentum and energy in collisions

❑ All moving objects have energy due to their motion. This form of energy is known as kinetic energy. Moving objects also have momentum.

○ In some collisions the kinetic energy before is equal to the kinetic energy after the collision. Such collisions are known as **elastic**.

○ In other collisions the kinetic energy after the collision is less than before. Such collisions are known as **inelastic**. The kinetic energy is converted to heat and sound during the collision.

○ The total energy is always conserved in both types of collision (see page 53).

total energy into system = total energy out of system

IGCSE Physics Summarised

Elastic collision

○ **Example 1**

A rubber ball A of mass 1.0 kg moving to the right at a velocity of 4.0 m/s collides with another rubber ball B of mass 1.0 kg, which is stationary.

(a) Ball A is stationary after the collision. Calculate the velocity of ball B.

(b) Show that the collision is elastic.

Before collision

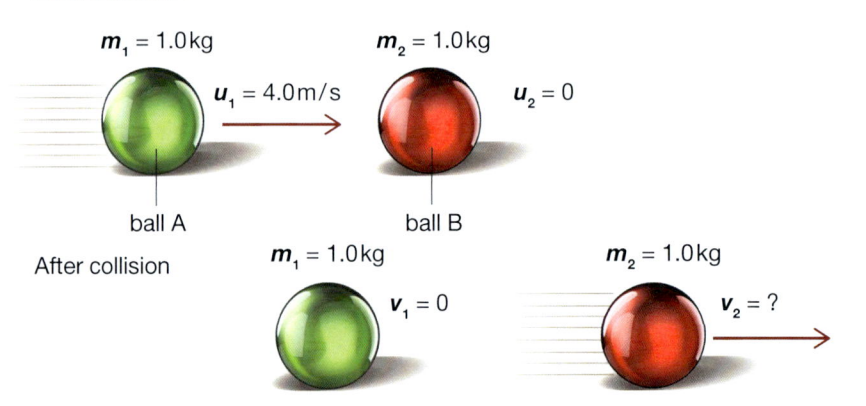

$m_1 = 1.0$ kg, $u_1 = 4.0$ m/s — ball A
$m_2 = 1.0$ kg, $u_2 = 0$ — ball B

After collision

$m_1 = 1.0$ kg, $v_1 = 0$ — ball A
$m_2 = 1.0$ kg, $v_2 = ?$ — ball B

(a) **Step 1** List all the information in symbol form and change into appropriate and consistent SI units if required.

Before collision:
$m_1 = 1.0$ kg
$u_1 = +4.0$ m/s
$m_2 = 1.0$ kg
$u_2 = 0$

After collision:
$m_1 = 1.0$ kg
$v_1 = 0$
$m_2 = 1.0$ kg
$v_2 = ?$

Step 2 Use the correct formula.

momentum before = momentum after

$$m_1 u_1 + m_2 u_2 = m_1 v_1 + m_2 v_2$$

Step 3 Calculate the answer by putting the numbers into the formula.

momentum before:

$$m_1 u_1 + m_2 u_2 = (1.0 \times 4.0) + (1.0 \times 0) = 4.0 \text{ kg m/s}$$

momentum after:

$$m_1 v_1 + m_2 v_2 = (1.0 \times 0) + (1.0 \times v_2)$$

momentum before = momentum after

$$4.0 = v_2$$

Ball B moves to the right at 4.0 m/s.

ALWAYS REMEMBER TO STATE THE UNIT FOR CALCULATED QUANTITIES.

(b) **Step 1** List all the information in symbol form and change into appropriate and consistent SI units if required.

All values before and after collision are as (a) but

$v_2 = +4.0 \, m/s$

Step 2 In an elastic collision, **k.e.** is conserved. Use the correct equation.
Total kinetic energy before collision:

$$\frac{1}{2} m_1 u_1^2 + \frac{1}{2} m_2 u_2^2$$

Total kinetic energy after collision:

$$\frac{1}{2} m_1 v_1^2 + \frac{1}{2} m_2 v_2^2$$

Step 3 Calculate the answer by putting the numbers into the equation.
Total kinetic energy before collision:

$$\frac{1}{2} \times 1.0 \times 4.0^2 + \frac{1}{2} \times 1.0 \times 0^2 = 8.0 \, J$$

Total kinetic energy after collision:

$$\frac{1}{2} \times 1.0 \times 0^2 + \frac{1}{2} \times 1.0 \times 4.0^2 = 8.0 \, J$$

There is no kinetic energy change in this collision, therefore the collision is elastic.

ALWAYS REMEMBER TO STATE THE UNIT FOR CALCULATED QUANTITIES.

Inelastic collision

○ **Example 2**

A toy car of mass 1.0 kg moving to the right at a velocity of 1.0 m/s collides with and sticks to a second toy car of mass 1.5 kg moving at a velocity of 2.0 m/s to the left.

(a) Calculate the velocity of the combined cars after the collision.

(b) Show that the collision is inelastic.

Before collision

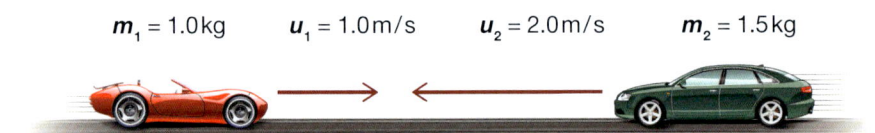

$m_1 = 1.0$ kg $u_1 = 1.0$ m/s $u_2 = 2.0$ m/s $m_2 = 1.5$ kg

After collision

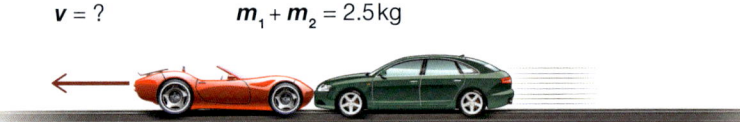

$v = ?$ $m_1 + m_2 = 2.5$ kg

(a) **Step 1** List all the information in symbol form and change into appropriate and consistent SI units if required.

Before collision: After collision:

$m_1 = 1.0$ kg
$u_1 = +1.0$ m/s $m_1 + m_2 = 2.5$ kg
$m_2 = 1.5$ kg $v = ?$
$u_2 = -2.0$ m/s

Since the toy cars stick together, the masses are combined after the collision.

Step 2 In all collisions with no external forces, momentum is conserved (see page 52). Use the correct formula.

momentum before = momentum after

$m_1 u_1 + m_2 u_2 = (m_1 + m_2) v$

Step 3   Calculate the answer by putting the numbers into the equation.

$$(1.0 \times 1.0) + (1.5 \times -2.0) = 2.5\,v$$
$$-2.0 = 2.5\,v$$
$$v = -0.80\,\text{m/s}$$

The cars move to the left at 0.80 m/s.

(b) **Step 1**   List all the information in symbol form and change into appropriate and consistent SI units if required.

All values before and after collision are as part (a) but
$v = -0.80\,\text{m/s}$

Step 2   In an inelastic collision, **k.e.** is not conserved. Use the correct formula.
Total kinetic energy before collision:

$$\frac{1}{2}m_1 u_1^2 + \frac{1}{2}m_2 u_2^2$$

Total kinetic energy after collision:

$$\frac{1}{2}(m_1 + m_2)v^2$$

Step 3   Calculate the answer by putting the numbers into the equation.

Total kinetic energy before collision:

$$\frac{1}{2}(1.0 \times 1.0^2) + \frac{1}{2}(1.5 \times [-2.0]^2) =$$
$$0.50 + 3.0 = 3.5\,\text{J}$$

Total kinetic energy after collision:

$$\frac{1}{2}(2.5 \times 0.80^2) = 0.80\,\text{J}$$

The kinetic energy decrease in the collision is:
$3.5 - 0.80 = 2.7\,\text{J}$

Therefore the collision is inelastic because the kinetic energy has decreased.

ALWAYS REMEMBER TO STATE THE UNIT FOR CALCULATED QUANTITIES.

Energy resources

- Energy cannot be created or destroyed; it can be **transformed** from one form into another. In other words, the useful output energy is decreased.

- Almost all our energy comes initially from the **Sun**. The exceptions are geothermal energy, energy from nuclear fission and tidal energy. The Sun's energy comes from nuclear fusion.

- **Fossil fuels (coal, oil** and **gas)** are currently the main source of energy used worldwide. The **chemical energy** in fossil fuels can be used to produce electricity.

- Fossil fuels are the very highly compressed remains of dead plants and animals that lived many millions of years ago. Fossil fuel reserves will eventually run out. They are a **finite** or **non-renewable** resource.

  Energy sources are **renewable** or **non-renewable**:

Renewable	Non-renewable
waves	oil
solar	coal
tidal	gas
wind	uranium (nuclear fission)
hydroelectric	
geothermal	

Nuclear fission and nuclear fusion

- **Nuclear fission**, which occurs in nuclear power plants, is the process in which large nuclei like uranium **split** into smaller ones, releasing a huge quantity of energy.

Unit 1 General physics

○ **Nuclear fusion**, which occurs in the Sun, is the process in which two small nuclei **combine** to form a larger one, again releasing an enormous amount of energy. In the Sun fusion occurs when hydrogen atoms combine to form helium (see page 230).

Generation of electricity from fossil and nuclear fuels

❏ **Chemical** energy from fuel is converted to **electrical** energy. There are essentially four main steps to generating electricity in a power station when using fossil fuels (coal, oil and gas) or nuclear fuel. After generation the electrical energy is transmitted all over the country (see page 225).

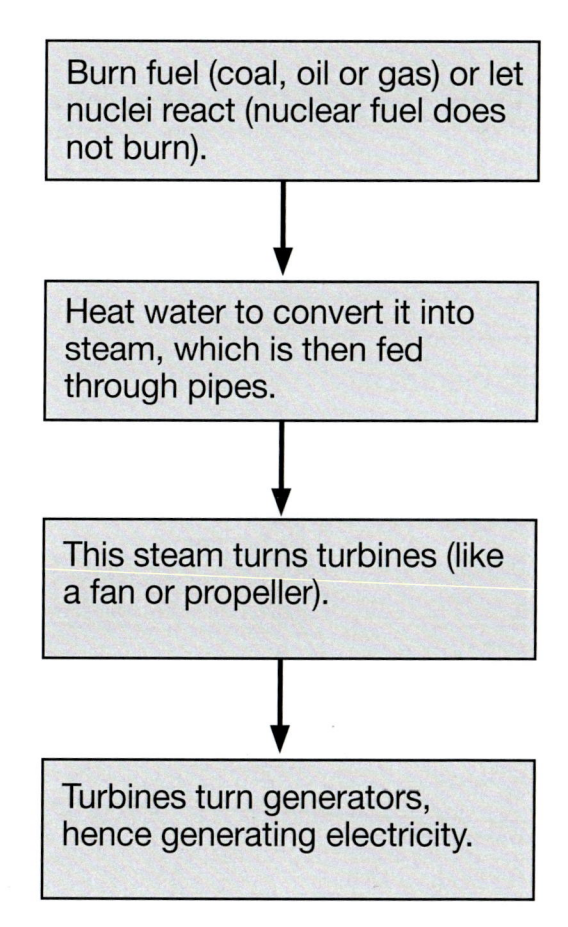

```
┌─────────────────────────────────┐
│ Burn fuel (coal, oil or gas) or let │
│ nuclei react (nuclear fuel does │
│ not burn).                      │
└─────────────────────────────────┘
                │
                ▼
┌─────────────────────────────────┐
│ Heat water to convert it into   │
│ steam, which is then fed        │
│ through pipes.                  │
└─────────────────────────────────┘
                │
                ▼
┌─────────────────────────────────┐
│ This steam turns turbines (like │
│ a fan or propeller).            │
└─────────────────────────────────┘
                │
                ▼
┌─────────────────────────────────┐
│ Turbines turn generators,       │
│ hence generating electricity.   │
└─────────────────────────────────┘
```

Generation of electricity from renewable sources

❑ ***Solar energy – makes use of energy from the Sun***
Solar cells, known as photovoltaic (PV) cells, convert light energy directly into electricity.

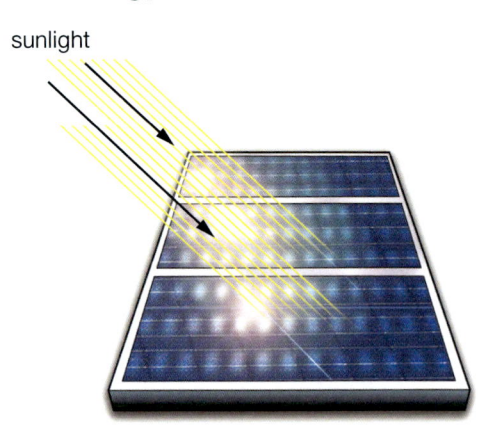

❑ ***Wind energy – uses wind to turn turbines***
Wind energy turns the turbine blades in a wind turbine. The turbines rotate the generator. The kinetic energy of the wind is converted into electrical energy by the generator.

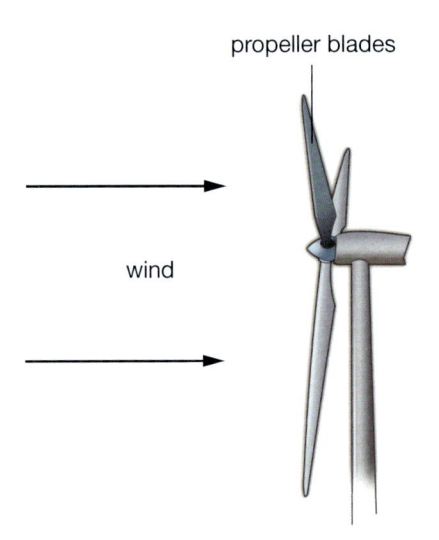

❑ ***Wave energy – makes use of ocean waves to turn turbines***
Generators are driven by the transverse (up and down) motion of the wave. The kinetic energy of the wave motion causes water to rise and fall in the air chamber. The air above the water causes the turbine to turn and electricity is produced by the generator.

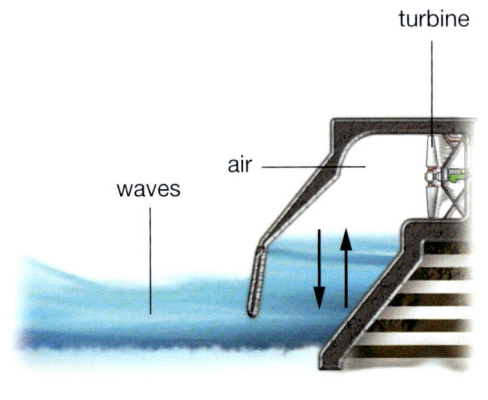

❏ **Tidal energy – makes use of incoming and outgoing tides**

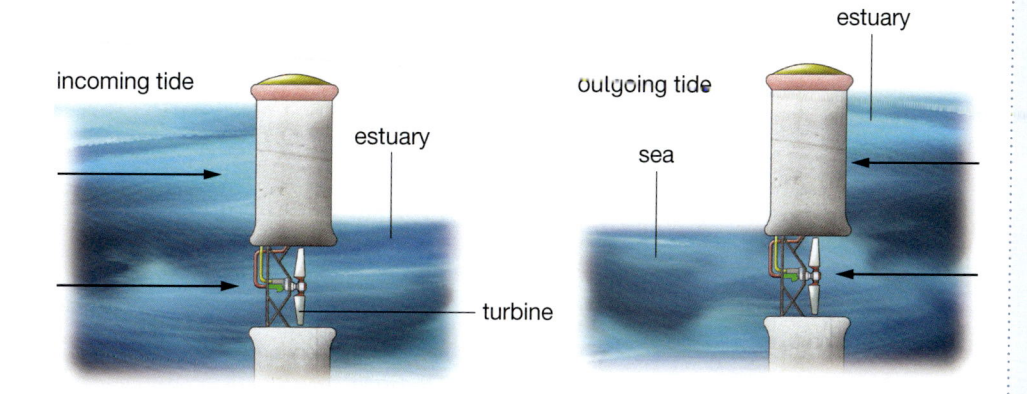

Tidal energy causes the water to move. As it does so it turns the turbines and electricity is generated. The gravitational potential energy and kinetic energy of the water converts to electrical energy.

❏ **Geothermal energy – uses heat energy from the Earth's core**

Water is fed down through pipes several kilometres underground, passing through hot rocks. These rocks heat the water until it turns into very hot steam. The steam is fed back up to ground level and is used to turn turbines and generate electricity.

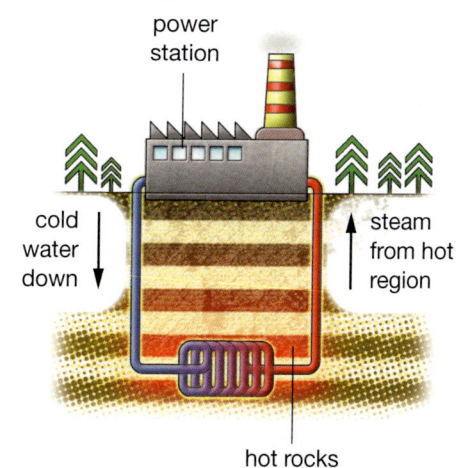

❏ **Hydroelectric energy – uses a dam to trap water**

A dam is built to trap water, sometimes in a valley where there is an existing lake. Water is allowed to flow through tunnels in the dam, to turn turbines and thus drive the generators that produce electricity.

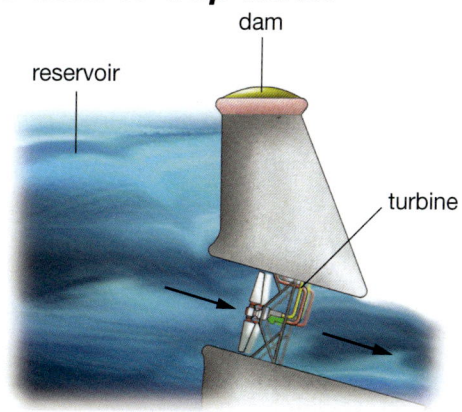

Other useful forms of energy obtained from these sources

❏ Electricity is a very useful form of energy for modern life. But the same natural energy sources have long provided us with other types of useful energy.

- Chemical energy from fuels such as wood or coal is transferred to heat energy by combustion (burning).
- Stored chemical potential energy in petrol can be used to move engines and vehicles (kinetic energy).
- Geothermal energy can be used to heat homes using a series of pumps and pipes.
- Solar energy can be used directly as heat and light energy. Solar water-heating panels can be fixed on roofs to capture this efficiently.
- Wind and watermills have been used for centuries, for example for pumping water, grinding flour and crushing rocks.

Advantages of non-renewable sources

❏ These fuels are far more energy dense than renewable sources and thus, in many countries, they allow production of larger amounts of energy in comparison to renewable sources. They are also relatively easy to transport and can be stored ready for use.

Disadvantages of non-renewable sources

❏ Fossil fuels are becoming increasingly expensive to mine (coal) and drill for (oil), as reserves are running out (being **depleted**). They cause **global warming** due to high carbon dioxide (CO_2) emissions when they are burnt. Some also produce sulfur dioxide (SO_2), which causes **acid rain**.

Nuclear power stations are expensive to build and any radiation leak or explosion may have a devastating effect on the immediate population and environment, as well as the population and environment at large.

Advantages of renewable sources

❏ They are all regarded as clean (producing little or no pollution) and will not run out. The fuel itself is cheap or free.

Disadvantages of renewable sources

- ❏ Generally all renewable energy sources have high initial installation costs and the supply may not be constant.

- ❏ *Solar energy*
    - When the sun doesn't shine (e.g. at night) no electricity is produced.
    - Dirty solar panels are inefficient.
    - To install panels is an expensive process.
    - The panels take up large areas.

- ❏ *Wind energy*
    - When the wind doesn't blow no electricity is produced.
    - Wind farms can destroy the natural beauty of a landscape and are noisy.
    - Offshore wind farms are expensive to build and need to be avoided by shipping.

- ❏ *Wave energy*
    - Waves vary in size and therefore will produce varying quantities (amounts) of energy. When the sea is calm, no electricity is produced.
    - Installation is expensive and challenging.
    - Boats would have to be careful and be aware of the location of the turbines to prevent accidents.

- ❏ *Tidal energy*
    - Many countries do not have suitable locations.
    - Might affect local marine life and destroy habitats.

- ❏ *Geothermal energy*
    - There are few locations that are suitable.
    - It is often necessary to drill very deep and this makes the energy very expensive to obtain.

- ❏ *Hydroelectric energy*
    - The local environment may be destroyed because water needs to be stored behind a dam, the land behind it flooding.
    - Dams are expensive to build.

❑ **Efficiency** is the term used to describe how good a device is at transferring one form of energy into another.

It is better for a car to be 70% rather than 50% efficient. An efficiency of 70% means the car is transferring seven-tenths of the chemical energy from the fuel into **useful energy**. The other 30% or three-tenths is transferred to less useful heat energy and sound energy. (Note that in cold countries some of this 30% would be useful energy rather than wasted as it would keep us warm in the car.)

Useful energy can be described as the type of energy we want from a device or what it's built (designed) to do. For example, a television set gives out light, heat and sound energy. We want to see and hear the television, so in this example light and sound are useful forms of energy. We do not need the heat; therefore it is described as wasted energy.

○ The formula for efficiency is:

$$\text{efficiency} = \frac{\text{useful energy out}}{\text{total energy in}} \times 100\%$$

> **Top Tip**
>
> **Remember:**
> total energy into system = total energy out of system

❑ For example, 100 J of electrical energy from a battery is converted into 100 J of energy in a torch.

If 15 J comes out as light then the other 85 J is wasted as heat. The efficiency of the torch is 15%.

If 10 J comes out as light then 90 J is wasted as heat. The efficiency of the torch in this case is 10%.

100 J of **input energy** is **always** converted into **100 J** of **output energy**.

Work

- **Work** is done when a force is exerted through a distance. More work is done when:
    - the **force** is **larger**
    - the **distance** moved is **greater**.

- The **work done** is a measure of the amount of energy transferred by the force; it has the symbol **W**, and like all forms of energy, its unit is the **joule** (J).

- Work done is related to the **force** and the **distance moved in the direction of the force** by the formula:

    $$W = Fd = \Delta E$$

    W = work done (J)
    F = force (N)
    d = distance moved by the force (m)
    ΔE = energy transferred (J)

- **Example**
  A 20 N force pushes a toy tractor and 50 J of energy is transferred. Calculate the distance moved.

    Step 1 List all the information in symbol form and change into appropriate and consistent SI units if required.

    $F = 20\,\text{N}$
    $W = \Delta E$ = energy transferred = 50 J
    $d = ?$

    Step 2 Use and rearrange the correct formula.

    $$W = Fd \implies d = \frac{W}{F}$$

    Step 3 Calculate the answer by putting the numbers into the formula.

    $$d = \frac{W}{F} = \frac{50}{20} = 2.5\,\text{m}$$

    ALWAYS REMEMBER TO STATE THE UNIT FOR CALCULATED QUANTITIES.

Power

- If a man pushes a weight through a distance, he does work. His power is related to how quickly he does that work; the faster he does it the more power he has.

- A car does work when its engine exerts a driving force and it moves through a distance. Cars with more powerful engines can do work quicker than less powerful ones. They can usually travel faster.

- **Power** is the rate of doing work. It has the symbol P, and its unit is the **watt** (W); one watt is defined as **one joule per second**:

  $1\,W = 1\,J/s$

  A 100 W lamp will use 100 J of energy every second.

  - Power is related to energy (work done) and time by the formula:

    $$P = \frac{W}{t} = \frac{\Delta E}{t}$$

    P = power (W)
    W = work done (J)
    ΔE = energy transferred (J)
    t = time (s)

  - It follows that **efficiency** can also be expressed by the

    $$\text{efficiency} = \frac{\text{useful power output}}{\text{power input}} \times 100\,\%$$

> **Note**
>
> The **time taken** to do something **does not** have an effect on the **work done**. The time taken **does** have an effect on the **power**.

○ **Example**

An athlete exerts an average force of 30 N while running. He runs a distance of 1.6 km in 5.0 minutes. Calculate his power.

Step 1 List all the information in symbol form and change into appropriate and consistent SI units if required.

$F = 30\,N$
$d = 1.6\,km = 1600\,m$
$t = 5.0\,minutes = 5.0 \times 60 = 300\,s$
$P = ?$

Step 2 Choose the correct formulae.

$W = Fd \qquad P = \dfrac{W}{t}$

Step 3 Calculate the answer by putting the numbers into the formulae.

$W = Fd = 30 \times 1600 = 48\,000\,J$

$P = \dfrac{W}{t} = \dfrac{48\,000}{300} = 160\,W$

ALWAYS REMEMBER TO STATE THE UNIT FOR CALCULATED QUANTITIES.

> **Top Tip**
>
> In some calculations, **two** equations have to be used to find the answer as shown above.

Section 1.8 Pressure

- Pressure is defined as the **force per unit area**.

- Pressure has the symbol **p** and its unit is the **pascal** (Pa).

- Pressure, force and area are related by the formula:

  $$p = \frac{F}{A}$$

  p = pressure (Pa) or (N/m²)
  F = force (N)
  A = area (m²)

- 1 Pa is equivalent to 1 N/m² (**newton per metre squared**).

- Pressure can be increased by **increasing the force** on a constant area.

- Pressure can be increased by **decreasing the area** for a constant force.

- A girl weighing 500 N and wearing high heels with an area of 2 cm² in contact with the floor can make indentations on a wooden floor, because the pressure she exerts is 2 500 000 Pa.

  An elephant weighing 40 000 N and standing on all four feet, a total area of 0.4 m², would exert a pressure of only 100 000 Pa.

- A sharp knife cuts bread more easily than a blunt knife. This is because the sharp knife has a much smaller surface area in contact with the bread than a blunt knife. When you push down on the knife (exert a force) the sharp knife exerts greater pressure on the bread and it cuts easily.

Example

A rectangular block has a mass of 200g and dimensions 5.0cm × 12.0cm × 2.0cm. Calculate the maximum pressure that can be exerted by the block on the surface on which it rests.

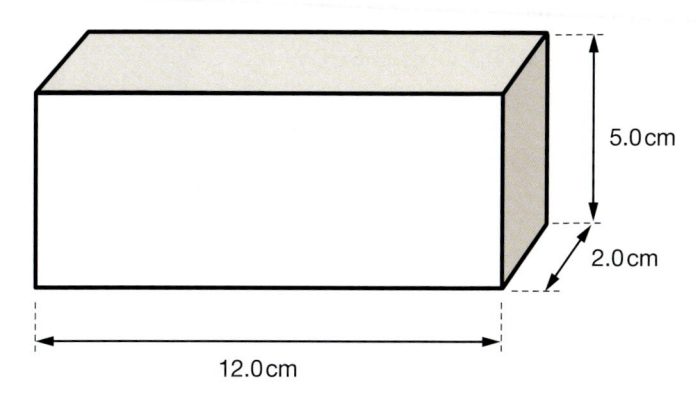

Step 1 List all the information in symbol form and change into appropriate and consistent SI units if required. The question asks for the maximum pressure. The maximum pressure is exerted when the block rests on the smallest area.

$m = 200g = 0.20 kg$
$A = 5.0 \times 2.0 = 10 cm^2 = 10 \times 10^{-4} m^2$
$= 1.0 \times 10^{-3} m^2$

$p = ?$

Remember: Take care when changing the units of area:

$1 cm^2 = 1 \times 10^{-4} m^2$

Step 2 Choose the correct formulae.

$F = \text{weight} = W = mg \qquad p = \dfrac{F}{A}$

Step 3 Calculate the answer by putting the numbers into the formulae.

$W = mg = 0.2 \times 10 = 2.0 N$

$p = \dfrac{F}{A} = \dfrac{2.0}{1.0 \times 10^{-3}} = 2000 Pa$

ALWAYS REMEMBER TO STATE THE UNIT FOR CALCULATED QUANTITIES.

More practical examples of pressure

- Snow shoes shaped like tennis rackets are used in very cold places so that people don't sink into the snow. The **large area** of the snow shoe **decreases** the **pressure** on the snow.

- Camels have **large feet** to **decrease** the **pressure** they exert on the ground so that they do not sink into the sand.

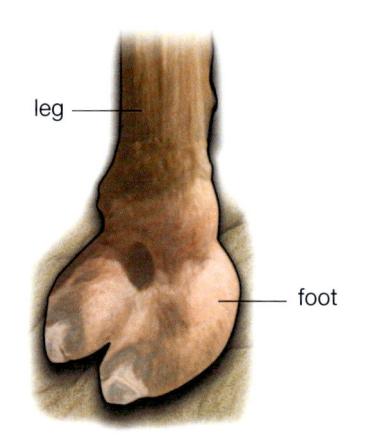

- Pushing a drawing pin into soft wood is easy as the point has a **very small area** so there is a **very large pressure**.

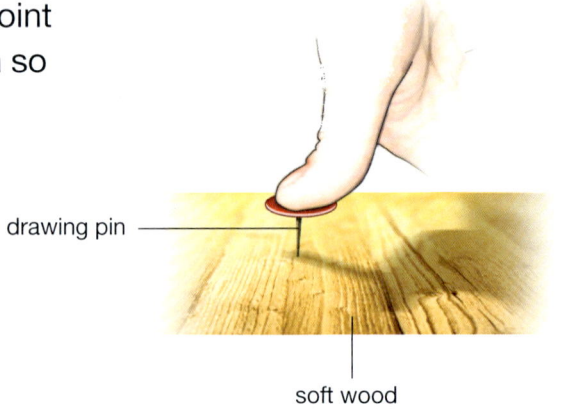

The mercury barometer

❏ The air around us exerts a pressure. This pressure is called **atmospheric pressure** and it may be measured using a **mercury barometer**. The mercury barometer is made of a glass tube, sealed at the top, with its lower end open in a dish of mercury.

There is a downwards force due to atmospheric pressure on the mercury in the dish at the bottom; this causes the mercury to rise up in the glass tube.

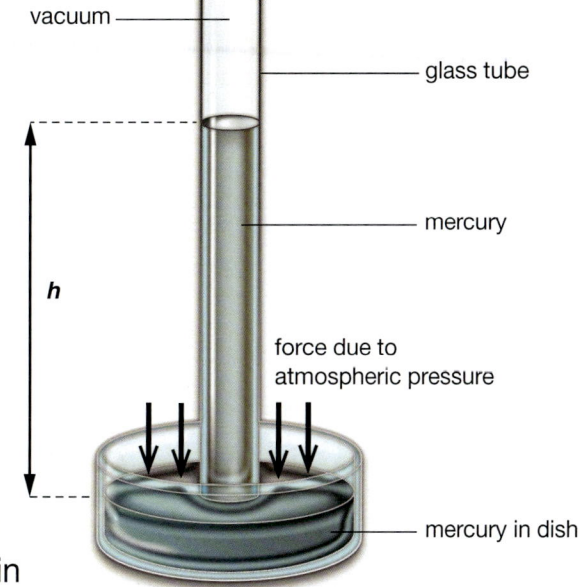

The space above the mercury is a **vacuum**. There is **no pressure** on the mercury at the top of the tube. The height h from the top of the mercury in the dish to the top of the mercury in the tube can be used to calculate **atmospheric pressure**.

The height h fluctuates about the value 760 mm as the atmospheric pressure fluctuates. Atmospheric pressure can be stated in millimetres of mercury (mmHg) so we can say that the atmospheric pressure is 760 mmHg.

❏ Atmospheric pressure has the following properties.

- Its effect **acts equally in all directions**.
- It **decreases** as altitude (height above sea level) increases because the air molecules become further apart (the air is less dense).
- It **increases** as altitude decreases because air molecules become closer together (the air is more dense). At **sea level**, it is approximately 760 mmHg which is equal to 100 000 Pa (100 kPa).

- ❏ Liquids also exert pressure. Pressure has the following properties in **liquids**:
  - The pressure acts **equally in all directions**; it pushes on every surface that the liquid is in contact with.
  - The pressure increases with **increasing depth** due to the weight of the column of liquid (height *h*) above the measurement point.
  - The pressure also depends on the density of the liquid; the greater the **density**, the greater the pressure at a given depth.

- ❏ The pressure of a liquid varies directly with the height *h* and the density *ρ* (rho) of the liquid. As a result the pressure will be greater at the bottom of a 10 cm tall measuring cylinder full of water than the pressure due to 10 cm of oil in an identical cylinder because the water is more **dense**.
  To achieve the same pressure with oil you would need a taller cylinder to increase the height *h*.

- ○ In a liquid, pressure, density, gravitational field strength and height are related by the formula:

$$p = \rho g h$$

p = pressure (Pa or N/m²)
ρ = density (kg/m³)
g = gravitational field strength (N/kg)
h = height (m)

Remember: $\text{density} = \dfrac{\text{mass}}{\text{volume}} \qquad \rho = \dfrac{m}{V}$

By rearranging the equation we see that the height depends on pressure, density and gravitational field strength:

$$h = \dfrac{p}{\rho g}$$

> **Top Tip**
>
> Mercury is used in a barometer instead of water because it is very dense. If we used water, the barometer would have to be a very long instrument.

○ **Example**
A diver is 18m below the surface of water of density 1000kg/m³. Calculate the pressure the water exerts on him.

Step 1 List all the information in symbol form and change into appropriate and consistent SI units if required.

g = 10N/kg
h = 18m
ρ = 1000kg/m³
p = ?

Step 2 Use the correct formula.

$p = \rho g h$

Step 3 Calculate the answer by putting the numbers into the formula.

$p = \rho g h$ = 1000 × 10 × 18 = 180 000 Pa
= 1.8 × 10⁵ Pa

ALWAYS REMEMBER TO STATE THE UNIT FOR CALCULATED QUANTITIES.

Top Tip

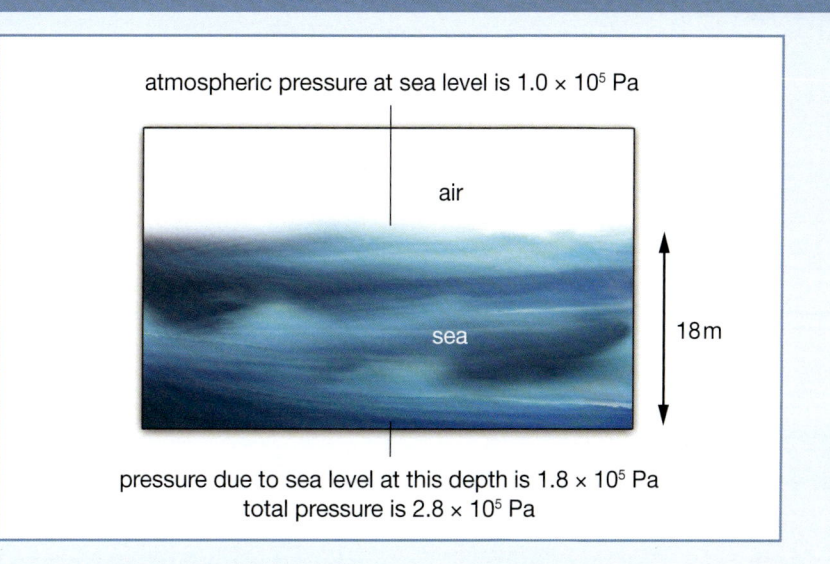

atmospheric pressure at sea level is 1.0 × 10⁵ Pa

air

sea   18m

pressure due to sea level at this depth is 1.8 × 10⁵ Pa
total pressure is 2.8 × 10⁵ Pa

The **total** pressure on the diver is 280 000 Pa (2.8 × 10⁵ Pa) rather than 180 000 Pa (1.8 × 10⁵ Pa) because atmospheric pressure of 100 000 Pa (1.0 × 10⁵ Pa) has to be added to the pressure caused by sea water.

The mercury manometer

❏ A **manometer** can be used to measure gas pressures. It consists of a tube of glass or plastic, bent into a U shape, and is partially filled with a liquid that is often oil or mercury. Side A is attached to the **gas supply**. The liquid moves if there is a pressure difference; the greater the pressure difference the bigger the height difference.

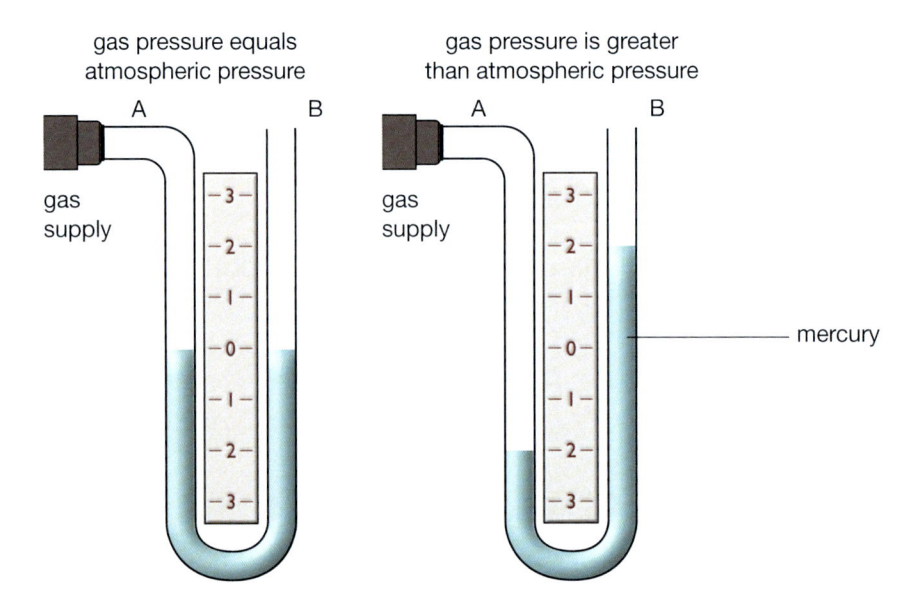

❏ The height difference tells you how much above atmospheric pressure the actual pressure of the gas is.

○ In the right-hand diagram, there is a difference in the level of mercury in the two sides of the tubes. On the left, the level is 2.0 cm below 0 and on the right, it is 2.0 cm above 0, a total difference of 4.0. Therefore the gas pressure is 4.0 cm of mercury, or more commonly 40 mm Hg above atmospheric pressure.

> **Top Tip**
>
> Atmospheric pressure is often expressed as 760 mm Hg. Gas pressure is sometimes expressed relative to this. For instance a pressure of 40 mm Hg above atmospheric means a total pressure of 40 + 760, or 800 mm Hg.

Unit 2 Thermal physics

Section 2.1 Simple kinetic molecular model of matter

States of matter

❑ The three states of matter are:
   1. solid
   2. liquid
   3. gas.

❑ The kinetic model of matter explains the behaviour of solids, liquids and gases in terms of how they are **arranged** and the **movement** of the particles (**molecules** or **atoms**) from which they are made.

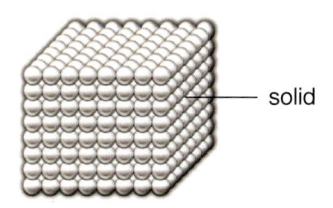

— solid

❑ **Properties of a solid**
   - The particles are fixed **close together** in a regular lattice pattern, for example arranged in neat rows.
   - The particles **vibrate** around a fixed position but do not move from place to place.
   - Solids have a **fixed shape**.
   - Solids have a **fixed volume** (provided the temperature and pressure remain constant).

○ There are very **strong forces of attraction** between particles because they are close together. (Individual particles cannot break free from the lattice.)

○ The solid cannot be squashed or compressed easily because the particles are close together and it takes a very large force to push them closer.

83

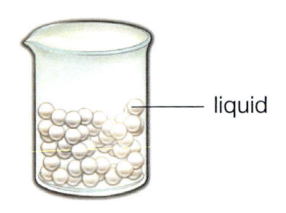

❑ **Properties of a liquid**

- The particles are still close together but have no fixed arrangement.
- The particles are **free to move** and tend to **slip and slide** past each other.
- Liquids do not have a fixed shape; they **take the shape** of the bottom of the container they are in.
- Liquids have a **fixed volume** (provided the temperature and pressure remain constant).

○ There are slightly **weaker forces of attraction** between particles.

○ Liquids cannot be squashed or compressed as the particles are close together.

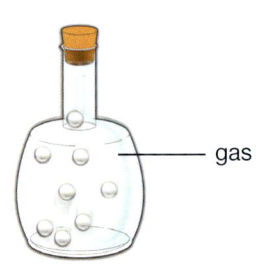

❑ **Properties of a gas**

- The particles are very much further apart than in liquids.
- The particles **move very fast** in **random directions**.
- The particles are constantly **colliding** with each other and the walls of the container.
- Gases have **no fixed shape** or **volume**; they fill any container in which they are placed.

○ Gases can be squashed or **compressed** because the particles are far apart.

○ There are **negligible** forces of attraction between particles.

Unit 2 Thermal physics

> **Note**
>
> The **temperature of a substance** is a measure of the **kinetic energy** of its particles. The **faster** the particles move or vibrate, the greater the temperature and the **hotter** the substance. The **slower** the particles move or vibrate, the **colder** the substance.

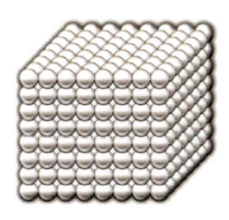

The table below lists the main properties of solids, liquids and gases.

Solid	Liquid	Gas
very slightly compressible	incompressible	can be compressed
fixed shape	takes on the shape of the bottom of the container	fills the container, taking on its shape
fixed volume	fixed volume	no fixed volume
static	can flow	can flow
particles very closely packed	particles disordered and closely packed	particles far apart
very strong forces of attraction between atoms/molecules	strong forces of attraction between atoms/molecules	almost no forces of attraction between atoms/molecules

> **Note**
>
> Particles in the solid state have the least energy and particles in the gaseous state have the most energy.
>
> **Remember:** When it comes to states of matter:
>
> **think movement – think energy**.

Molecular model

Brownian motion

❑ The Brownian motion experiment uses a microscope to view very small smoke particles in a transparent air cell. It shows the smoke particles moving randomly. This is evidence of free-moving air molecules.

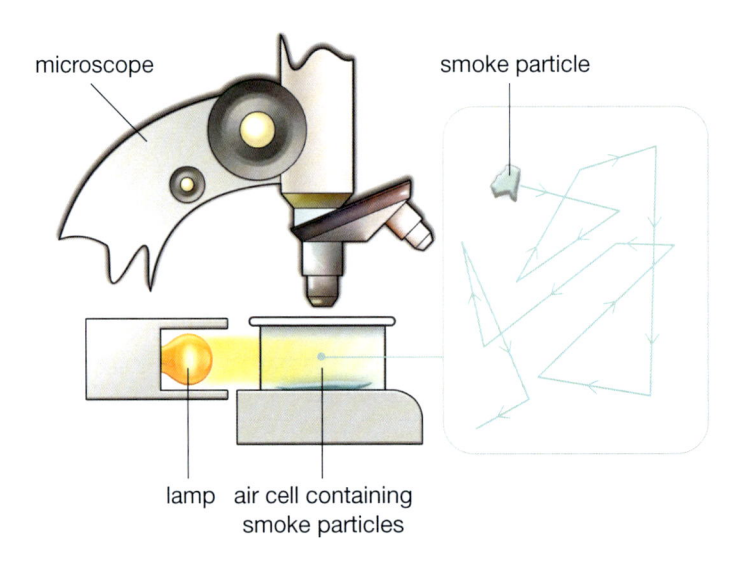

❑ According to the kinetic theory, a gas such as air is made up of an extremely large number of tiny, invisible molecules that have relatively large spaces between them and are constantly moving randomly.

❑ These molecules have no effect on each other except when they collide. All **collisions** are **elastic**; this means the molecules bounce off each other without losing kinetic energy.

❑ The smoke particles in the air cell are constantly bombarded from all sides by the air molecules. The larger smoke particles can be seen to move in **random straight lines** in a zig-zag pattern (see diagram above).

○ This random movement is due to **collisions** between the **very light** but **fast-moving air molecules**, which are too small to be seen, and the **smoke particles**, which are larger and can be seen.

Pressure caused by gases

- All **moving particles** have **kinetic energy** and **momentum**.

- **Gas** molecules have **large** energies (because they are moving very fast). The molecules in a gas do not all have the same energy; some have more than others. The **average** kinetic energy is related to the **temperature** of the gas.

- **Increasing** the **temperature** of a gas increases the **average kinetic energy** of particles within that gas and they move **faster**.

- The gas particles exert a **force** on the walls of a container when they collide with it. The **pressure** exerted by the gas particles is the **force per unit area**. The total pressure of the gas is the effect of the sum of all the collisions with the wall.

- When the molecules collide with the walls of the container they change direction and bounce back. This means that the velocity (a vector quantity) has changed. If the velocity changes, the momentum changes, and force is related to a change of momentum (see page 50).

- The force is **larger** if the particles are moving **faster** (because the change in momentum is greater) or if there are **more particles** colliding with the walls **per second**.

- A **higher temperature** causes **more** collisions in a given time and causes the collisions to be **harder** because the particles are moving faster. Consequently, a higher temperature causes a **larger pressure**.

> **Top Tip**
>
> As the **temperature** of a gas **increases**, the **pressure increases** because:
> - the molecules or particles have **greater kinetic energy**
> - they hit the walls of the container **harder**
> - they hit the walls of the container **more often**.

Evaporation

❑ Evaporation involves a change of state from liquid to vapour. (A vapour is a substance in its gaseous state below its boiling point – see page 110.) Evaporation is the reason why wet clothes dry on a washing line, or a saucer of water eventually dries up.

❑ Evaporation is the **escape** of the most **energetic molecules** from the **surface** of a liquid. Not all molecules in a liquid have the same energy. It is those molecules that are moving the fastest (i.e. those with the greatest kinetic energy) and are at the surface that can escape.

❑ If the more energetic molecules leave the liquid, it follows that the average energy of the molecules of the liquid falls and so its temperature falls. Evaporation **leads to the cooling** of a liquid.

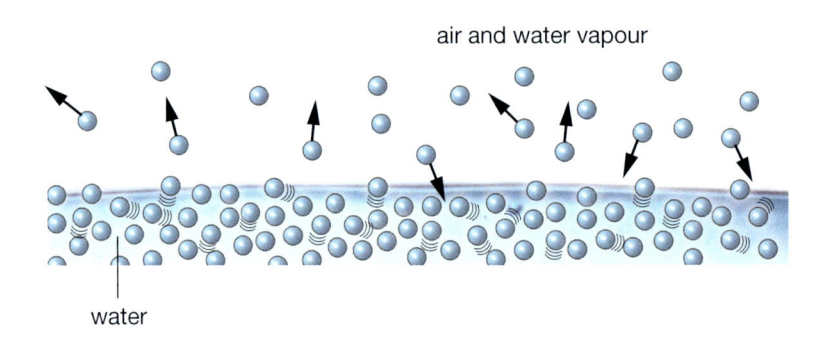

[Diagram: air and water vapour above water surface, with molecules escaping]

○ Some vapour molecules return to the surface of the liquid unless they are removed by draught or wind.

Factors affecting the rate of evaporation

○ A **higher temperature increases** the average kinetic energy of the liquid molecules; therefore many more molecules have a chance of escaping from the surface.

○ A **larger surface area** allows more molecules to be closer to the surface, which increases the rate of evaporation. (This is like making an entrance to a school wider, allowing more people to get into the school at one time.)

- Draught or wind movement can take away those molecules that have escaped the liquid's surface, so that these molecules cannot return to the liquid's surface.

- In a humid atmosphere there are already a lot of water molecules in the air. These molecules can join the liquid water without difficulty. Consequently, when water evaporates into a humid atmosphere, some water molecules return to the liquid's surface as fast as others escape. (An analogy that is useful here is that you are digging a hole and someone else is filling it in behind you.)

Cooling effect of evaporation

- Rapid (fast) evaporation has a noticeable cooling effect on objects in contact with the liquid.
    - As the liquid evaporates, it takes away **thermal energy** from an object. (Human sweating is an example of this.)
    - Evaporation has a cooling effect because molecules with the greatest kinetic energy escape; consequently the molecules left behind have less kinetic energy and therefore the temperature falls. The liquid then draws heat energy from a body with which it is in contact.

Boiling and evaporation

- Boiling (see page 110) is also a change of state from liquid to vapour (vaporisation). But boiling and evaporation differ in the following ways.
    - Boiling occurs **throughout** the liquid, whereas evaporation only occurs at the **surface** of the liquid.
    - Boiling only occurs at **one temperature** (100°C for water at standard atmospheric pressure), whereas evaporation occurs at **all temperatures**.
    - Boiling requires a supply of energy for the liquid to reach the required temperature, whereas evaporation occurs at all temperatures, utilising the internal energy of the liquid.
    - Boiling is usually a rapid process. Evaporation is usually a slow process but can be speeded up by the presence of a draught over the surface of the liquid and an increased surface area in contact with the air.

Pressure changes in a gas

❏ Important variables that can vary for a gas are:
   1. pressure
   2. volume
   3. temperature
   4. mass.

Boyle's Law

❏ Boyle's Law explains how the pressure and volume are related when the **temperature is kept constant**.

❏ If the piston of a bicycle pump with a sealed end is pushed in at constant temperature, then the further you push it in, the harder it gets to push.

❏ This is because the **pressure** inside the container (pump) **increases as the volume decreases**.

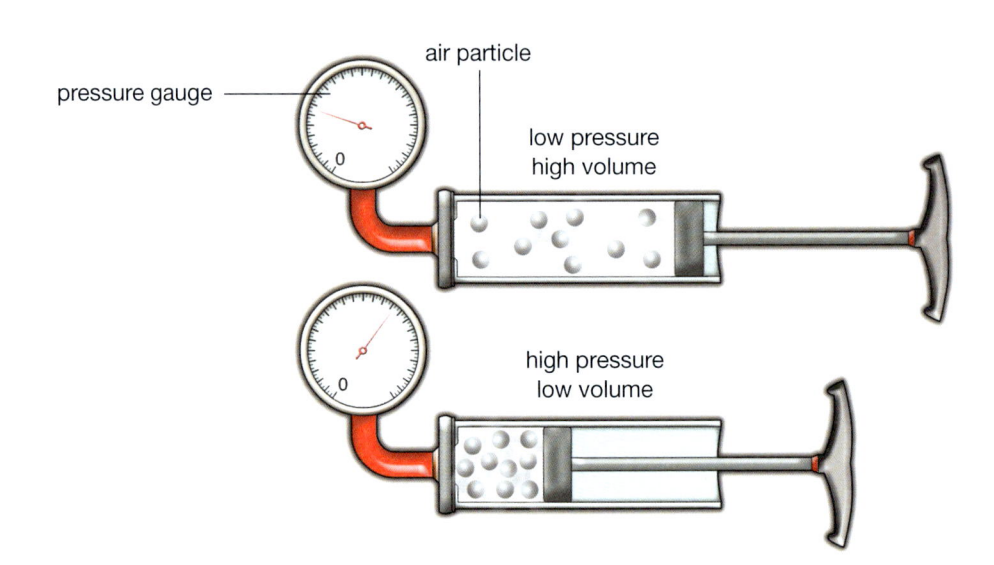

❏ The pressure increases because the gas particles are squeezed into a smaller space. Therefore there are more collisions per second amongst the particles and with the container walls. The more frequent collisions cause a larger force on the walls and hence a larger pressure. The collisions do not get harder because the temperature does not change; it is the frequency of collisions that is responsible for the change in pressure.

> **Top Tip**
>
> **When the volume of a fixed mass of gas is decreased at a constant temperature, the pressure increases.**
> This is because the molecules or particles are squeezed into a smaller space and therefore **collide more often** with the walls of the container and so the pressure increases. They **do not** move faster.

○ Boyle's Law states that:

For a fixed mass of gas at constant temperature, the pressure is inversely proportional to the volume.

○ This means that for a fixed mass of gas at constant temperature, the **pressure** multiplied by the **volume** is **constant**.

$p \times V$ = constant

○ In other words, when **the pressure increases**, **the volume decreases** and vice versa, as shown in the graph below. Notice that the line does not touch either axis.

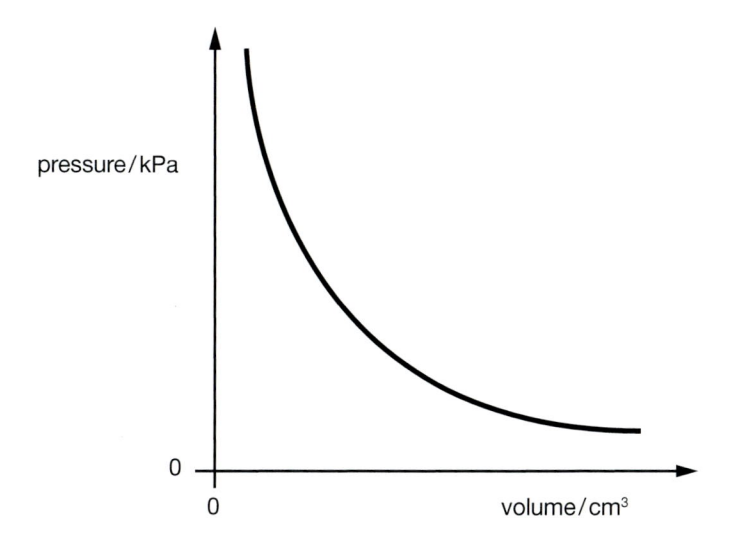

○ Pressure and volume are related by the formula:

$$p_1 \times V_1 = p_2 \times V_2$$

p_1 = initial pressure (Pa or N/m²)
V_1 = initial volume (cm³ or m³)
p_2 = final pressure (Pa or N/m²)
V_2 = final volume (cm³ or m³)

○ ### Example
A bicycle pump contains 300 cm³ of air at atmospheric pressure. The air is compressed slowly. Calculate the pressure when the volume of the air is compressed to 125 cm³. (Remember that atmospheric pressure is 100 kPa.)

Step 1   List all the information in symbol form and change into appropriate and consistent SI units if required.

p_1 = 100 kPa = 100 000 Pa
V_1 = 300 cm³
V_2 = 125 cm³
p_2 = ?

Step 2   Use and rearrange the correct formula. Because the air is compressed slowly we can assume the temperature remains constant and apply Boyle's Law.

$$p_1 \times V_1 = p_2 \times V_2$$

Therefore

$$p_2 = \frac{p_1 \times V_1}{V_2}$$

Step 3   Calculate the answer by putting the numbers into the formula.

$$p_2 = \frac{p_1 \times V_1}{V_2} = \frac{100\,000 \times 300}{125} = 240\,000\,\text{Pa}$$

ALWAYS REMEMBER TO STATE THE UNIT FOR CALCULATED QUANTITIES.

> **Top Tip**
>
> When a gas is in a cylinder of constant cross-sectional area, the volume is proportional to the length of the cylinder occupied by the gas. In this case, the formula
>
> $p_1 \times V_1 = p_2 \times V_2$
>
> can be replaced by
>
> $p_1 \times l_1 = p_2 \times l_2$
>
> where
>
> p_1 = initial pressure (Pa or N/m²) l_1 = initial length (cm and m)
> p_2 = final pressure (Pa or N/m²) l_2 = final length (cm and m)

Increasing the temperature of a gas

- [ ] The pressure of a **fixed volume** of gas is related to its temperature. As the temperature increases, so does the pressure.

- [ ] When the temperature of a gas is increased the **internal energy** of its molecules **increases**. The molecules move faster because the average kinetic energy of the molecules increases.

- [ ] The pressure of the gas is determined by the force of the particles colliding with the walls of the container and by the rate of collision.

- [ ] As the temperature increases, the force increases and so the **pressure of a gas increases with temperature**.

- [ ] As the volume of the container is kept constant, there will also be more molecules hitting the walls of the container every second and so the pressure is increased.

> **Top Tip**
>
> **When the temperature of a fixed mass of gas is increased, its pressure increases if the volume is kept constant.** This is because the kinetic energy and therefore average velocity of the molecules increases and so the force exerted on the walls of the container is greater.

Section 2.2 Thermal properties and temperature

Thermal expansion of solids, liquids and gases

❏ **Thermal expansion** is the **increase in volume** of a solid, liquid or gas that is caused by **heating**.

○ For the same temperature increase **gases** expand the most, then **liquids** and then **solids**.

Remember: The molecular separation in liquids is very slightly greater than in solids. The molecular separation in gases is much greater than in solids or liquids.

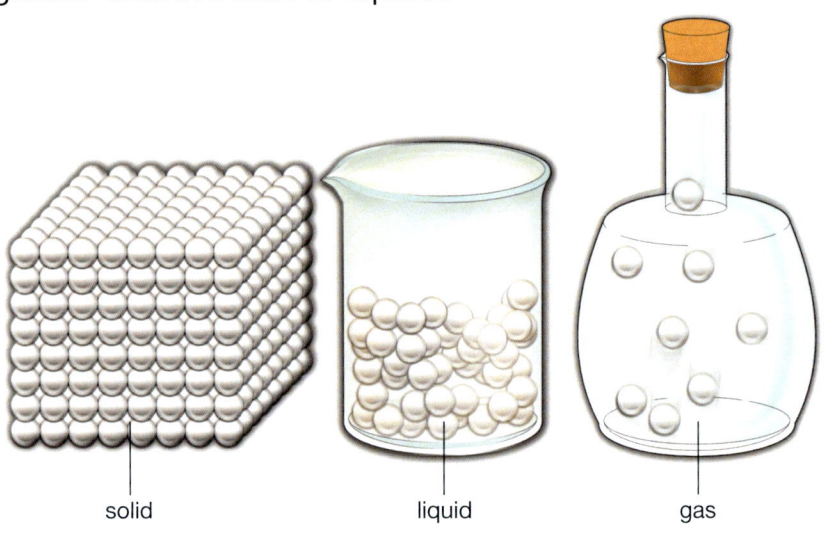

solid liquid gas

Expansion in solids

❏ When a solid substance is heated its particles vibrate more. This causes the particles to move slightly further apart and therefore the substance **expands** in all directions.

❏ When the substance is cooled its particles vibrate less. This causes the particles to move closer together and therefore the substance **contracts** in all directions.

> **Note**
>
> The atoms or molecules themselves **do not** expand individually; they move further apart collectively.
>
> A substance expands when its atoms or molecules move further apart in all directions due to an increase in their kinetic energy.

Unit 2   Thermal physics

Applications and consequences

There are several applications and consequences of expansion in solids. Some examples are given below.

❑   ***Applications – the bimetallic strip***
Bimetallic strips are made up of two metals such as brass and steel joined side by side. Brass expands more than steel when heated and contracts more when cooled down. This causes the strips to bend as shown below.

brass
steel

brass and steel
at room temperature

brass and steel
heated above
room temperature

brass and steel
cooled below
room temperature

The bimetallic strip is found in many devices such as fire alarms, electrical thermostats and old-fashioned car indicator lights.

❑   ***Consequences – overhead cables and bridges***
Overhead power cables like those shown below are left slack in the summer. This is because the cables contract in winter. If they became too tight, they could snap.

Metal joints as seen below can be found in many of the world's bridges. They allow room for thermal expansion and contraction.

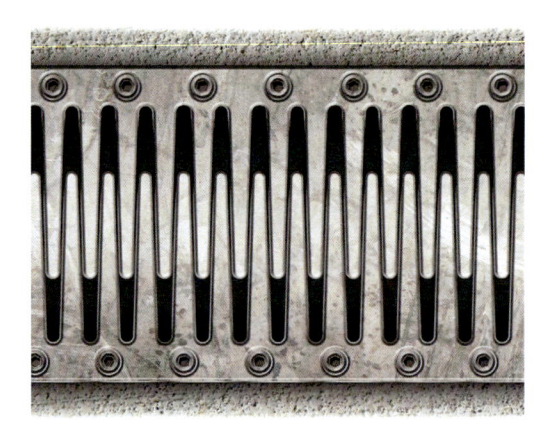

Expansion in liquids

- ❏ Most **liquids expand** when they get hotter. They expand **more than solids** but **less than gases**.

- ○ When a liquid is heated the kinetic energy of the particles increases and their separation increases. The particles in liquids have weaker forces of attraction compared to solids, so liquids expand more than solids when heated.

- ❏ Water is an exception; it behaves differently.

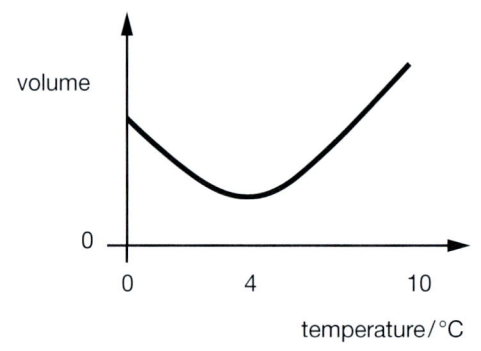

    - As the temperature decreases from 10°C to 4°C the volume decreases and density increases. The density of water is greatest at 4°C.
    - As the temperature decreases from 4°C to 0°C the volume increases and the density decreases.
    - At 0°C water freezes and turns to ice. Unusually water expands when it freezes; hence ice is less dense than water. This is why ice floats in water.
    - Most other substances are denser when they are in the solid state.

Unit 2 Thermal physics

Expansion in gases

○ The particles in gases are far apart and have negligible forces of attraction, so gases expand the most.

○ In the diagram below the 10N weight keeps the pressure constant, but allows the volume to vary. As the temperature increases the volume increases.

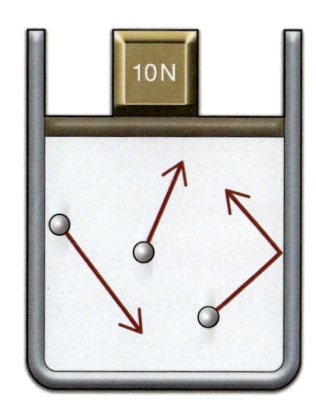

gas at low temperature

gas at high temperature

> **Top Tip**
>
> **When the temperature of a gas is increased at constant pressure** (shown in the diagram above), **the volume increases.**
>
> This is because the gas molecules have more **kinetic energy** and therefore they move **further apart**.

Remember: We do not need to allow a gas to expand if it gets hotter. A gas can be put inside a sealed container, so that it has a fixed volume.

❑ If a gas is heated in a sealed container its pressure increases (see page 93). If it increases too much there will be an **explosion**.

IGCSE Physics Summarised

Measurement of temperature

- Temperature is the measure of how **hot** an object is. The hotness of an object is a measure of the **kinetic energy** of the molecules that make up that object.

- A **thermometer** monitors a suitable physical property. In the case of a liquid-in-glass thermometer, it monitors volume of the liquid. Any device that includes a substance that changes uniformly with temperature can be calibrated and be made into a useful thermometer.

- To **calibrate** (create a scale) a thermometer we need known **upper** and **lower** fixed points (such as the steam point at 100°C and the ice point at 0°C). The **Celsius scale** was devised by Anders Celsius.

- There are several types of thermometer including the **mercury-in-glass** or the **alcohol-in-glass** thermometer, shown in the diagram below. This is a very common type of thermometer. The volume of the liquid and hence the length of the liquid column changes uniformly with temperature.

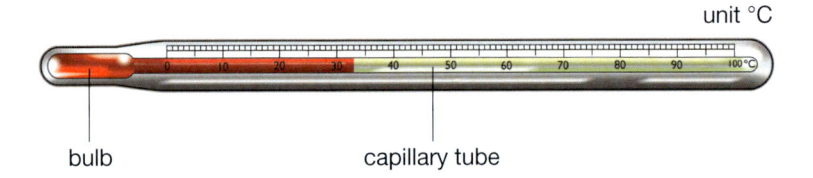

- **Mercury** is a liquid metal and as such is useful for measuring higher temperatures. Its boiling point is 357°C and it freezes at −39°C.

- **Alcohol** has a very low freezing point and as such is useful for measuring low temperatures. It has a boiling point of 78°C and freezes at −114°C.

- Other types of thermometers include:
  - The liquid crystal thermometer – it contains heat-sensitive liquid crystals that change colour to indicate different temperatures.

Unit 2 Thermal physics

- The constant-volume gas thermometer – it contains a gas; when the temperature rises, the pressure increases. The change in pressure is used to indicate a change in temperature.

○ The **range** of a thermometer is the difference between the **maximum** and **minimum** temperatures that the thermometer can read. A common range for liquid-in-glass thermometers is from −10°C to 110°C.

○ The range of a liquid-in-glass thermometer is limited by the length of the thermometer and can be increased by:

- increasing the diameter of the capillary – this means that the liquid will not expand as far along the tube per degree rise in temperature
- decreasing the volume of the bulb – this means there is less liquid and so it will not travel as far along the capillary tube as it is heated.

○ A liquid-in-glass thermometer is said to be **linear** if the liquid expands by the same amount for every degree Celsius rise in temperature. This means that the scale will be marked in degrees of equal size (as below).

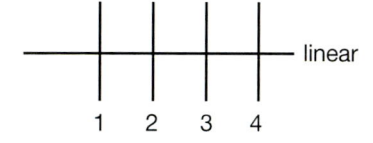

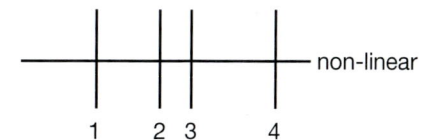

○ If the liquid did not expand uniformly, the scale would be non-linear. It would have to be marked with degrees of differing sizes (as above). It would be very difficult to calibrate and to use.

> **Top Tip**
>
> Characteristics that make a substance suitable for use in a liquid-in-glass thermometer include:
>
> - expanding uniformly over a large temperature range
> - a low specific heat capacity for quick response
> - a low freezing and high boiling point so that the substance remains liquid over a good range of temperatures.

○ A **sensitive** thermometer gives a **large response** to a **small change** in temperature. Alcohol is sometimes used for this purpose as it expands approximately six times as much per degree Celsius rise in temperature as mercury.

To increase the sensitivity in a liquid-in-glass thermometer, the **capillary tube** could be made **narrower** or the **bulb** could be made **larger** to give a larger volume of the liquid, so that the same change in temperature causes the liquid to move further along the capillary tube (see diagram page 98).

> *Top Tip*
>
> Thermometers can be made more sensitive by:
> - making the capillary tube narrower
> - making the bulb bigger
> - using a liquid that has a greater expansion per unit change in temperature.

❑ ***Example***
The syringe shown in the diagram below is sealed at one end. The syringe has air trapped inside it. Describe how you would calibrate the syringe so that its volume can be used to determine temperatures between 0 °C and 100 °C.

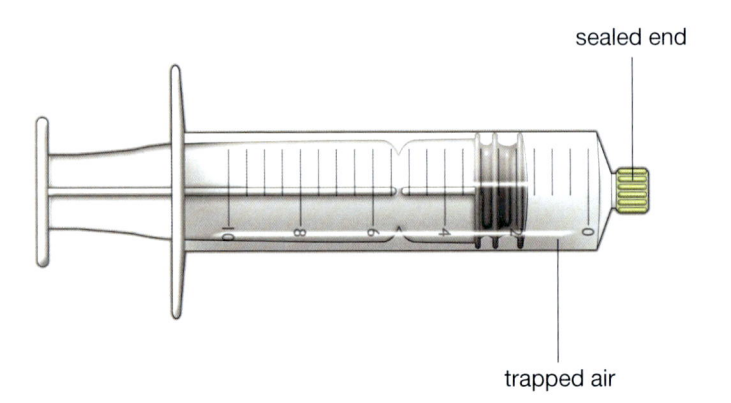

Step 1 Place the syringe in **pure melting ice** and wait for several minutes. Mark the position of the right-hand end of the piston **0 °C**.

Step 2 Place the syringe in **pure boiling water** and wait again. Mark the new position **100 °C**.

Unit 2 Thermal physics   Notes

Step 3 Measure the distance between the marked points and divide it by 10. Each division will represent 10 degrees Celsius.

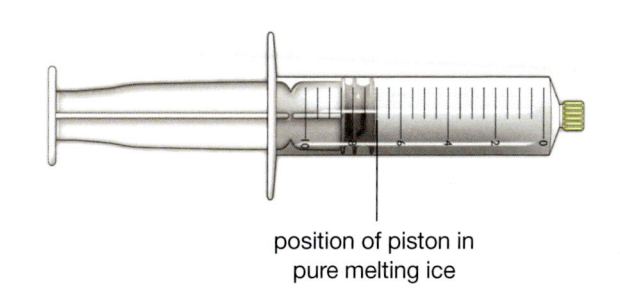

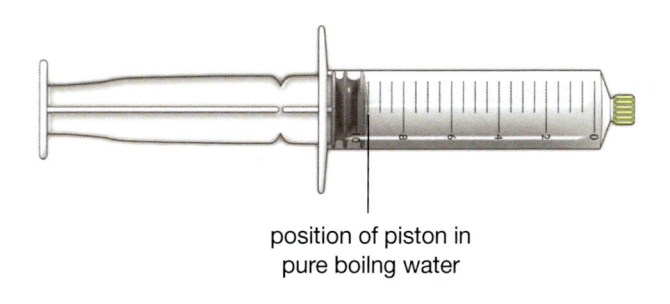

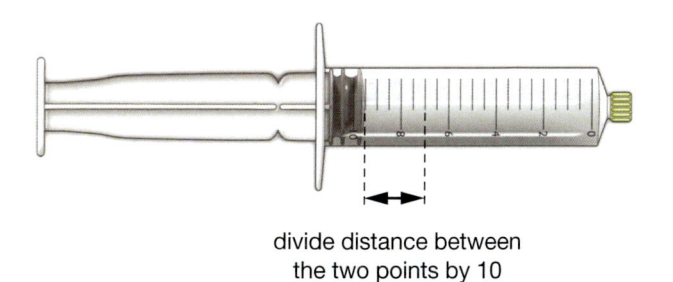

position of piston in pure melting ice

position of piston in pure boilng water

divide distance between the two points by 10

Step 4 The position of the piston now records the temperature against the scale you have marked. Notice that the gas expands inside the syringe when heated and contracts when it cools, as expected.

The thermocouple

- The thermocouple is an **electrical** thermometer consisting of wires of two different materials, e.g. copper and iron, joined together. It is the most common type used in industry. It can measure temperatures of over **1000°C** and is cheap to make.

- The thermocouple functions because two metals in contact with one another generate a tiny **voltage**. If there are two junctions, e.g. copper–iron–copper, the voltage produced varies with the **temperature difference between the two junctions**.

- Since it is electrical, a **remote dial** can be used to read the temperature. Consequently, temperatures can be measured in hard-to-reach or very hot places.

- Thermocouples **respond very quickly** to changes in temperature because they have a **low thermal capacity** (see page 104), and so are useful in measuring rapidly varying temperatures.

- *Using a thermocouple to measure temperature*
  The diagram shows an arrangement in which a thermocouple is being used to measure the temperature of a bunsen flame.

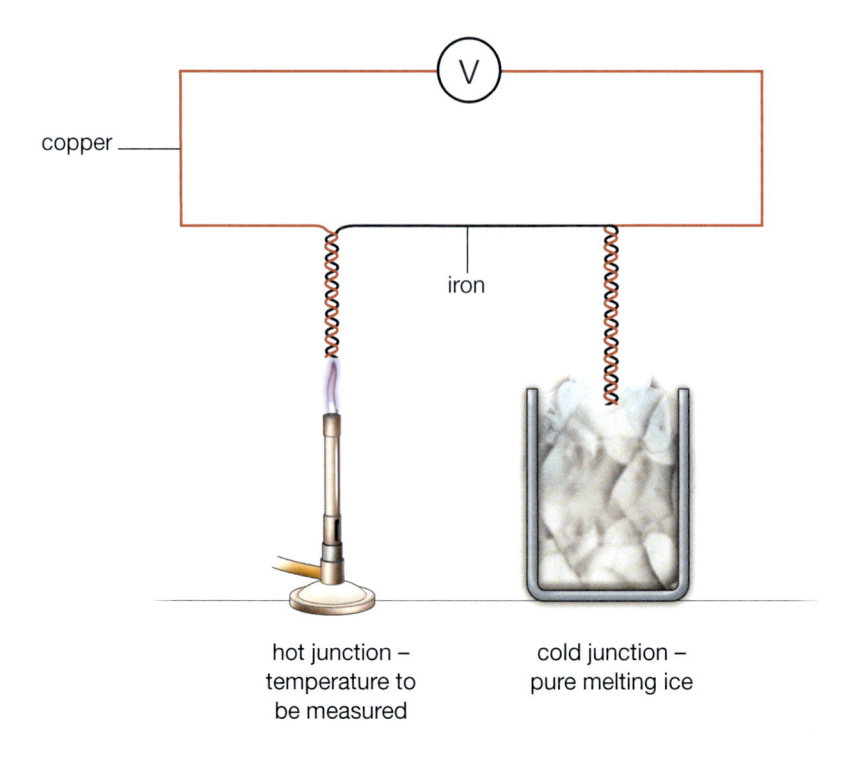

hot junction – temperature to be measured

cold junction – pure melting ice

- The copper wires of the thermocouple are connected to a sensitive voltmeter or galvanometer. The iron wire is twisted together with the copper wire to form junctions.

- One junction is placed in **pure melting ice** and the other is in contact with the substance at an **unknown temperature**.

- A sensitive meter allows very small voltages to be measured, so the thermocouple is a sensitive temperature measuring device.

How it works

- The voltmeter is **pre-calibrated** between two fixed points. The voltage can be converted into degrees Celsius using a calibration chart.

- If the junctions are at the same temperature the voltmeter will read zero because it is designed to respond to the temperature difference between these two points.

- If the temperatures of the junctions are different, then the voltmeter displays a voltage. The larger the temperature difference, the bigger the voltage.

> **Top Tip**
>
> A **temperature difference** between junctions causes a very small voltage between the wires. This voltage is measured using a sensitive voltmeter or galvanometer. The reading **changes with temperature**. Both junctions have to be at **known fixed points** for **calibration**.

Thermal capacity

- **Heat** is a form of energy (**thermal energy**) and its unit is the **joule** (J). When substances **absorb** thermal energy, their temperature usually rises.

- This rise in temperature shows that the **internal energy** of the substance is increasing. The internal energy is the sum of the potential and kinetic energies of the molecules.

- The **thermal capacity** (or heat capacity) **C** of an object is the amount of thermal energy needed to raise the temperature of the object by **1°C**. Its unit is the **joule per degree Celsius** (J/°C). Thermal capacity depends on the **mass** and the **substance** the object is made from.

- **Specific heat capacity** (s.h.c.) is the amount of thermal energy needed to raise the temperature of **1kg** of a substance by **1°C**. It has the symbol **c** and its unit is the **joule per kilogram degree Celsius**, J/(kg°C).

- The thermal capacity and the specific heat capacity are related by the following formula:

$$\text{thermal capacity} = C = mc$$

C = thermal capacity (J/°C)
m = mass (kg)
c = specific heat capacity (J/(kg°C))

> **Top Tip**
>
> On a hot summer's day in Spain the sand on a beach is very hot to step on; it is cold to step on at night. Sand has a **low specific heat capacity** and so it heats up and cools down quickly.

- The heat energy required to raise the temperature of a substance without changing state is:

$$E = cm\Delta\theta$$

E = heat energy (J)
c = specific heat capacity (J/(kg°C))
m = mass (kg)
$\Delta\theta$ = change in temperature (°C)

Example

The specific heat capacity of water is 4200 J/(kg °C).
Calculate:

(a) the thermal capacity of 3.0 kg of water

(b) the temperature change when 150 kJ of energy is given to 3.0 kg of water.

(a) **Step 1** List all the information in symbol form and change into appropriate and consistent SI units if required.

$$E = 150 \text{ kJ} = 150\,000 \text{ J}$$
$$c = 4200 \text{ J/(kg °C)}$$
$$m = 3.0 \text{ kg}$$
$$\Delta\theta = ?$$

Step 2 Use and rearrange the correct formula.

$$C = mc$$

Step 3 Calculate the answer by putting the numbers into the formula.

$$C = 3.0 \times 4200 = 12\,600 \text{ J/°C}$$

(b) **Step 1** As for part (a)

Step 2 Use and rearrange the correct formula.

$$E = cm\Delta\theta \implies \Delta\theta = \frac{E}{cm}$$

Step 3 Calculate the answer by putting the numbers into the formula.

$$\Delta\theta = \frac{E}{cm} = \frac{150\,000}{4200 \times 3.0} = 11.9 \text{ °C}$$

ALWAYS REMEMBER TO STATE THE UNIT FOR CALCULATED QUANTITIES.

Determining specific heat capacity

○ Remember that specific heat capacity is defined as the thermal energy needed to raise the temperature of 1 kg of a substance by 1 °C.

○ The following experiment describes how to determine the specific heat capacity of a metal.

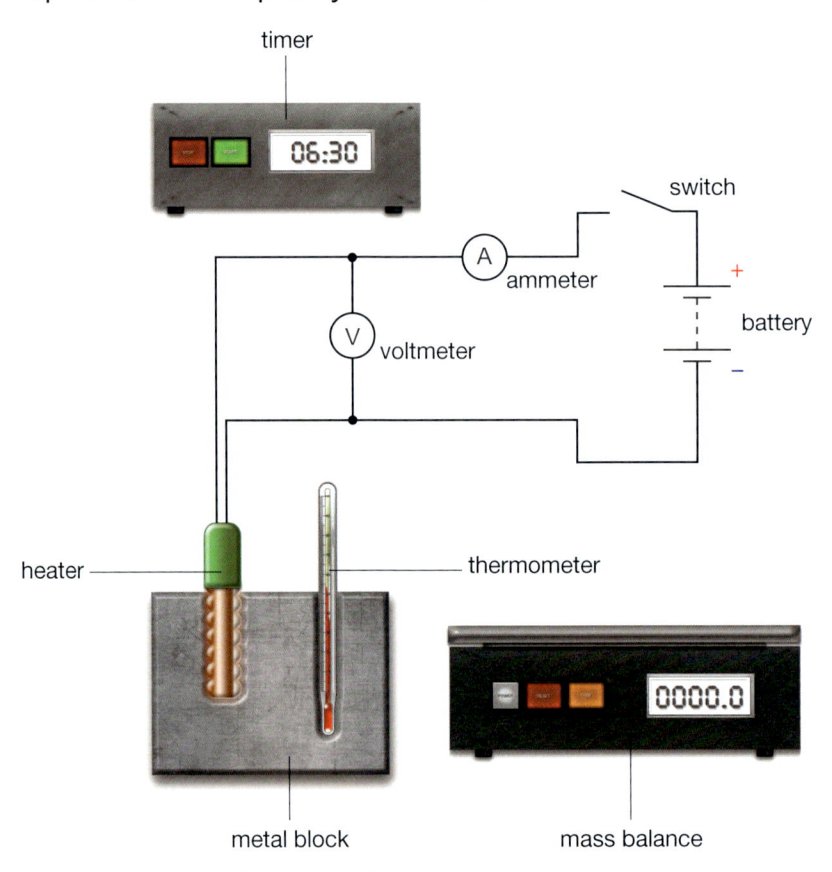

1. Set up the experiment as shown.

2. Place the metal block on the mass balance; record its mass m.

3. Record the initial temperature $\theta_{initial}$ of the block.

4. Switch the apparatus on, start the timer, and record the current and voltage readings.

5. After a time t switch off the power supply and record the final temperature θ_{final} of the block.

6. In the diagram the digital timer reads 6 minutes 30 seconds. This must be converted to seconds, i.e. 390 seconds.

Calculations to be made
1. Energy supplied by heater:

    $E = Pt$ where electrical power $P = IV$ (see *Top Tip* below)

    So, on substitution, $E = IVt$
2. The change in temperature:

    $\Delta\theta = \theta_{final} - \theta_{initial}$
3. The specific heat capacity of the metal block:

    $c = \dfrac{E}{m\Delta\theta}$

Improvements to the experiment
On occasion you may be asked to suggest improvements for practical experiments. In this case we are trying to find how thermal energy affects the block, therefore we want to make sure **minimal thermal energy is dissipated to the surroundings**.

The experiment could have been improved by:
1. adding lagging (insulation) on the sides, top and underneath the block
2. repeating the experiment to take an average.

Assumptions in the experiment and sources of error
We assume that all the heat energy supplied is used to heat the metal block. Some energy, however, is **dissipated** to the surroundings.

> **Top Tip**
>
> Energy supplied can be found using:
>
> **energy = power × time** or $E = Pt$
>
> The unit of energy is the joule (J), the unit of power is the watt (W) and the unit of time is the second (s).
>
> As **electrical power** = IV, then $E = IVt$
>
> The unit of energy is the joule (J), the unit of current is the ampere (A), the unit of voltage is the volt (V) and the unit of time is the second (s).

Calculating heat dissipated to the surroundings

○ Very often in calculations we assume that no heat is dissipated to the surroundings and that all input energy is transformed into useful output energy. Provided that the insulation is sufficiently effective, the heat that is lost to the surroundings is very small and can often be ignored.

○ The example below shows how to calculate heat dissipated to the surroundings, when this is not negligible.

Remember: Energy is conserved:
energy into system = energy out of system

Useful formulae are:

$$E = Pt$$
$$E = IVt$$

E = energy (J)
P = power (W)
I = current (A)
V = voltage (V)
t = time (s)

○ **Example**

A 1.0 kW immersion heater takes 15 minutes to raise the temperature of 1.0 kg of water by 26 °C. The specific heat capacity of water is 4200 J/(kg °C). Calculate the heat that is dissipated to the surroundings.

Step 1   List all the information in symbol form and change into appropriate and consistent SI units if required.

P = 1.0 kW = 1000 W
t = 15 minutes = 15 × 60 = 900 s
c = 4200 J/(kg °C)
m = 1.0 kg
$\Delta\theta$ = 26 °C
heat dissipated = ?

Step 2   Choose the correct formulae.

$$E = Pt$$
$$E = cm\Delta\theta$$

Step 3 Calculate the answer by putting the numbers into the formulae.

Heat energy given to the water by heater:

$E_{in} = Pt = 1000 \times 900 = 900\,000\,J$

Energy used to raise temperature of water (the rest is dissipated):

$E_{out} = cm\Delta\theta = 4200 \times 1.0 \times 26 = 109\,200\,J$

Energy dissipated to surroundings:

$E_{in} - E_{out} = 900\,000 - 109\,200 = 790\,800\,J$
$= 791\,000\,J$ (to 3 significant figures)
$= 791\,kJ$

ALWAYS REMEMBER TO STATE THE UNIT FOR CALCULATED QUANTITIES.

Top Tip

Very often the specific heat value obtained in calculations using experimental results is higher than the actual value; this is because heat is **dissipated to the surroundings**.

Note

If 1 kg of lead is given the same amount of heat as 1 kg of copper then the temperature rise of the lead is about **three times greater** than that of copper. Lead's specific heat capacity is about **three times smaller** than copper's.

Melting and boiling

- ❑ The three states of matter are **solid**, **liquid** and **gas**.

- ❑ The **changes of state** that can take place are:

    boiling – a liquid changing to a gas
    condensation – a gas changing to a liquid
    solidification – a liquid changing to a solid
    melting – a solid changing to a liquid

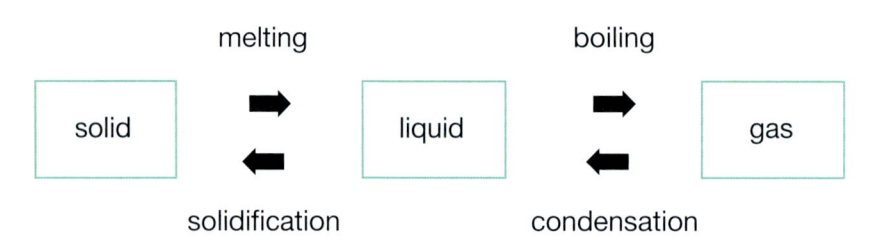

    ```
              melting              boiling
    ┌───────┐  →   ┌───────┐  →   ┌───────┐
    │ solid │      │ liquid│      │  gas  │
    └───────┘  ←   └───────┘  ←   └───────┘
           solidification      condensation
    ```

- ❑ Energy must be **provided** for melting or boiling.
    Energy is **given out** during solidification or condensation.

- ❑ The temperature of a substance remains **constant** whilst **changing state** at its melting point or boiling point. The temperature remains the same until the change of state is complete.

- ❑ The **melting point** is the temperature at which a solid turns into a liquid. The melting point of pure ice is 0 °C.

- ❑ The **boiling point** is the temperature at which a liquid turns into a gas. The boiling point of pure water is 100 °C.

- ❑ When melting or boiling, all the supplied energy is being used to **weaken** or **break** the bonds between molecules.

- ❑ When condensation occurs the molecules slow down and the bonds are **strengthened** or **formed**. These bonds bring the molecules closer together and the substance becomes liquid.

- ❑ When solidification occurs the temperature drops and fewer molecules have enough kinetic energy to overcome neighbouring attractions. The molecules can only vibrate, the substance gains a shape of its own and becomes a solid.

Unit 2 Thermal physics

❏ The amount of energy needed to change the state of a substance depends only on the **mass** and the **type of substance**. Different substances require different amounts of energy when boiling or melting.

○ The energy required to change a substance from a **solid** to a **liquid** at its melting point (without change of temperature) is the **latent heat of fusion**. (Fusion is an old word for melting.)

○ The energy required to change a substance from a **liquid** to a **gas** at its boiling point (without change of temperature) is the **latent heat of vaporisation**.

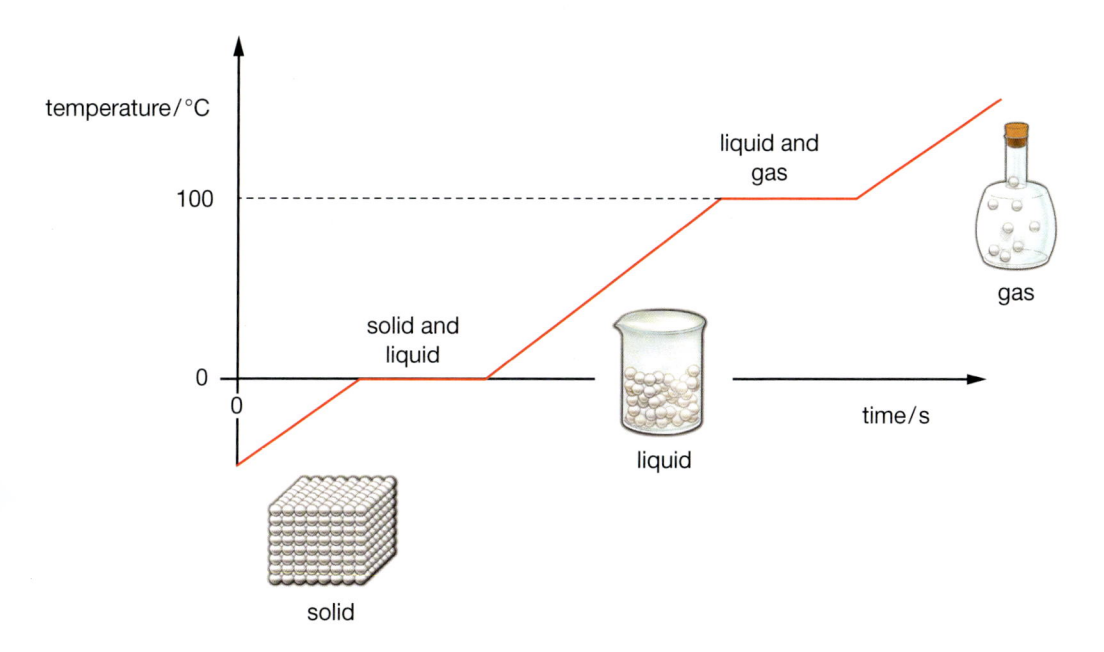

○ The graph above shows how the temperature changes with time when ice is heated.

- Whilst the ice is solid, its temperature increases as heat energy is supplied.
- When the ice reaches 0 °C the temperature remains constant as all the supplied heat energy is used to change the state from solid (ice) to liquid (water).
- The water changes temperature from 0 °C to 100 °C as heat energy continues to be supplied.
- At 100 °C the temperature remains constant as the water is turned to steam.

- The **specific latent heat of vaporisation** (l_v) is the energy needed to break molecules completely free from **1kg** of a liquid to form a gas.

- The **specific latent heat of fusion** (l_f) is the energy needed to **weaken** the bonds holding molecules in **1kg** of a solid to form a liquid.

  $l_{vaporisation}$ is always bigger than l_{fusion} for any given substance.

> **Top Tip**
>
> The **temperature does not increase** whilst a substance is **changing state** from solid to liquid or from liquid to gas because all the energy is being used to overcome the **forces of attraction** between molecules.

- For a substance to **change state**, the heat energy required is:

  $$E = ml$$

  E = heat energy (J)
  m = mass (kg)
  l = specific latent heat (J/kg)

- This applies to both changes of state, whether it be **fusion** (melting) or **vaporisation** (boiling).

- When a substance cools down and changes state from gas to liquid, or from liquid to solid, the latent heat is given out to the surroundings.

Example

A heater with a voltage of 24V and supplying a current of 6.0A is switched on for 5 minutes and 20 seconds and melts crushed ice. The volume of water melted is 137 cm³. Calculate the specific latent heat of fusion of ice.

Step 1 List all the information in symbol form and change into appropriate and consistent SI units if required.

V = 24V
I = 6.0A
m = 137 g = 0.137 kg
(1.0 cm³ of water has a mass of 1.0 g)
t = 5 minutes and 20 seconds
= (5 × 60) + 20 = 320 s
l = ?

Step 2 Choose and rearrange the correct formulae.

$E = Pt = IVt$ $l = \dfrac{E}{m}$

Step 3 Calculate the answer by putting the numbers into the formulae.

$E = Pt = IVt = 6.0 \times 24 \times 320 = 46\,080$ J

$l = \dfrac{E}{m} = \dfrac{46\,080}{0.137} = 336\,350$ J/kg

= 340 000 J/kg (to 2 significant figures)
= 340 kJ/kg

ALWAYS REMEMBER TO STATE THE UNIT FOR CALCULATED QUANTITIES.

Determining specific latent heat of fusion

○ Remember that the specific latent heat of fusion is defined as the energy needed to change 1 kg of a solid into a liquid without a change in temperature.

○ The following experiment describes how to determine the specific latent heat of fusion of ice.

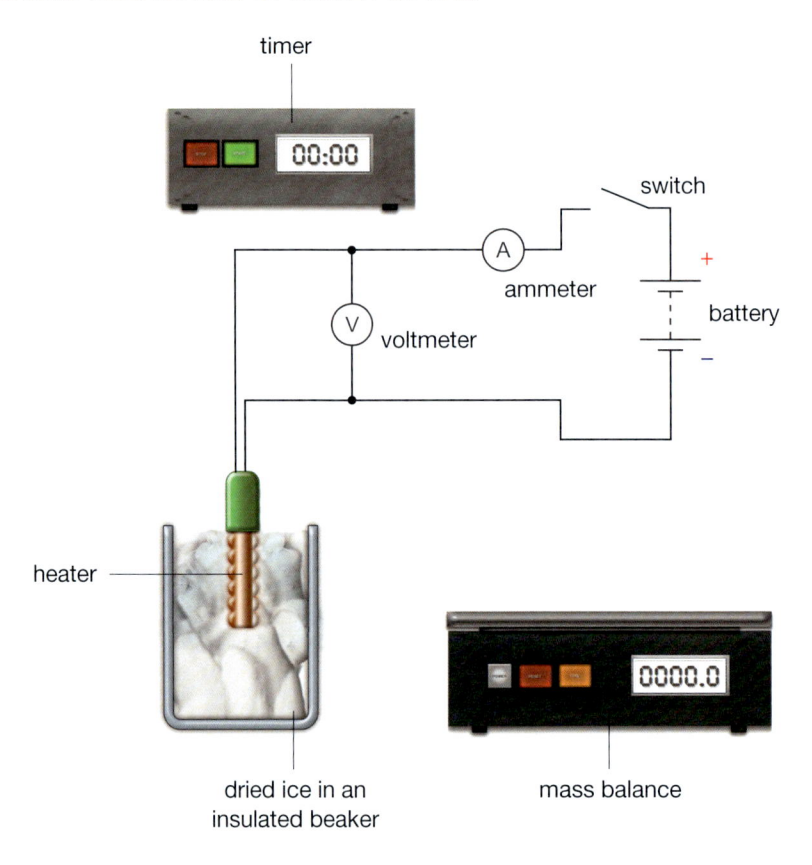

1. Set up the heater circuit as shown.

2. Measure the mass of the empty beaker on the mass balance.

3. Collect some ice and ensure it is at 0 °C by noting that the outside of the ice is just starting to melt.

4. Dry the ice on a paper towel and put a suitable amount into the beaker.

5. Put the heater into the beaker, switch it on and start the timer. Note the readings on the ammeter and voltmeter.

6. When the ice is almost completely melted, stir the water gently.

7. When all the ice has melted, switch off the heater and stop the timer.

8. Measure the mass of the beaker plus water.

Calculations to be made

1. Energy supplied by heater:

   $E = Pt$ where electrical power $P = IV$

   So, on substitution, $E = IVt$

2. The mass of ice melted:

   $mass_{ice\ melted} = mass_{beaker+water} - mass_{beaker}$

3. The specific latent heat of fusion:

   $l = \dfrac{E}{mass_{ice\ melted}}$

Improvements to the experiment

The experiment could have been improved by:

1. adding a plastic or polystyrene lid to the beaker
2. crushing the ice so there is less air between the ice and heater.

Assumptions in the experiment and sources of error

We assume that all the supplied heat energy is used to melt the ice at 0°C. In reality, some energy will be **dissipated** to the beaker and surroundings, and then some of the melted ice will warm slightly above 0°C.

> **Top Tip**
>
> Very often the specific latent heat value obtained differs from the actual value; this is because **heat** is **dissipated** to the surroundings from the heater. The ice itself will also **gain** heat from the surroundings.
>
> Using crushed ice will improve experimental results as there is less air between the ice, thus allowing better contact with the heater.

Determining specific latent heat of vaporisation

○ Remember that the specific latent heat of vaporisation is defined as the energy needed to change 1 kg of a liquid into a gas without a change in temperature.

○ The following experiment describes how to determine the specific latent heat of vaporisation.

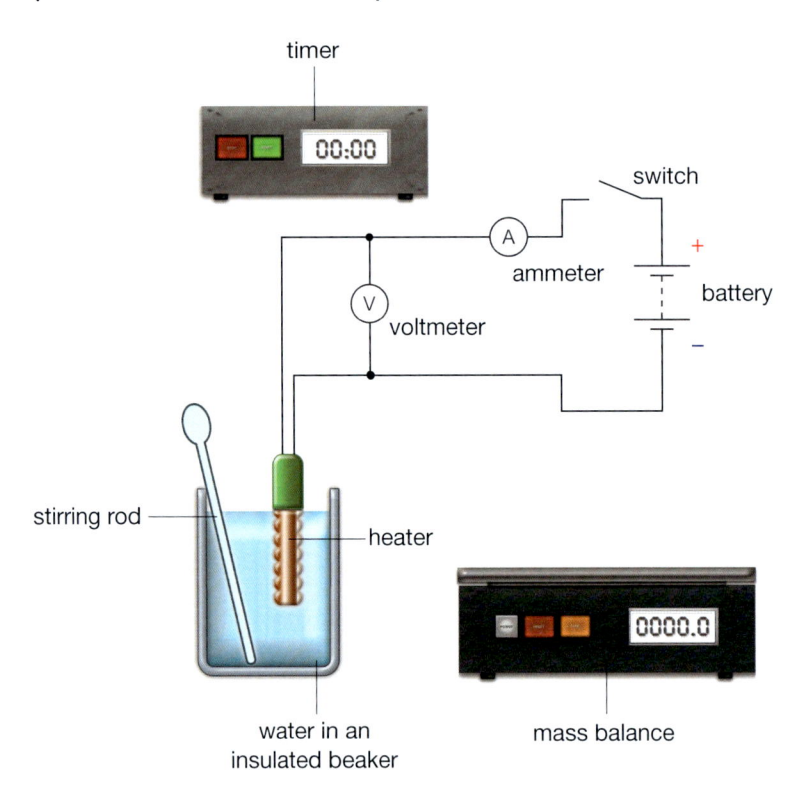

1. Put sufficient water into the beaker and measure the mass of the beaker and the water.

2. Put the heater into the beaker and switch it on. Note the reading on the ammeter and voltmeter.

3. Use a stirring rod to gently stir the water.

4. When the water starts to boil, start the timer.

5. Let the water boil for a suitable time. Switch off the heater and stop the timer.

6. Remove the heater.

7. Let the beaker cool to a safe temperature and measure the mass of the beaker and the remaining water.

Calculations to be made

1. Energy supplied by heater:

    $E = Pt$ where electrical power $P = IV$

    So, on substitution, $E = IVt$

2. The mass of water boiled off:

    $mass_{beaker + water\ before\ experiment} = m_{before}$
    $mass_{beaker + water\ after\ experiment} = m_{after}$
    $mass_{water\ boiled\ off} = m_{before} - m_{after}$

3. The specific latent heat of vaporisation:

    $$l = \frac{E}{mass_{water\ boiled\ off}}$$

Improvements to the experiment

The experiment could have been improved by creating a draught (use a small fan) over the beaker to ensure particles that have escaped from the liquid don't return to the surface.

Assumptions in the experiment and sources of error

We assume that all the heat energy supplied is used to heat the water. In reality, some energy will be **dissipated** to the beaker and surroundings.

Section 2.3 Thermal processes

Heat transfer

- The three principal methods of heat transfer are:
    1. conduction
    2. convection
    3. radiation.

- **Conduction** – Particles **gain energy** when heated. In a solid the particles cannot change positions but the particles vibrate, and this transfers the heat energy through the solid from particle to particle as they collide. In addition metals are particularly good conductors because they have **free-moving electrons**. The vibrating particles hit electrons and send them through the metal. This causes energy to be transferred quickly.

- **Convection** – A liquid or gas **expands** as it is heated and it becomes **less dense**; this causes the hot fluid to **rise** and the cooler fluid above it to **fall** in a circular fashion. The rising of the hot gas or liquid sets up a **convection current**.

    Remember: It is **NOT** heat that rises; it is the molecules of hot liquid or gas that rise.

- **Radiation** – Energy in the form of **infra-red radiation** (part of the electromagnetic spectrum) travels in all directions from any **hot body**. This is how heat energy is transferred through a vacuum, such as from the Sun through space to Earth.

- The amount of heat radiated from a body depends on its temperature; a very hot body radiates more heat than a cool body. A body with a large surface area will radiate more than a body of small surface area at the same temperature.

- Because infra-red radiation is an electromagnetic wave it can be reflected. This means that a highly polished surface will reflect heat radiation away, e.g. a polished metal plate behind the element of an electric fire reflects the heat into the room.

Unit 2 Thermal physics   Notes

> **Note**
>
> **Conduction** is the transfer of energy by vibrating particles in **solids**. Good conductors such as metals also have free-moving electrons to carry the energy.
>
> **Convection** is the transfer of energy by **hot gas or liquid rising** and **cold gas or liquid falling**.
>
> **Radiation** is the transfer of energy as an electromagnetic wave. Radiation is the only way thermal energy can travel through a vacuum.

Experiments demonstrating heat transfer

❏ **Conduction**

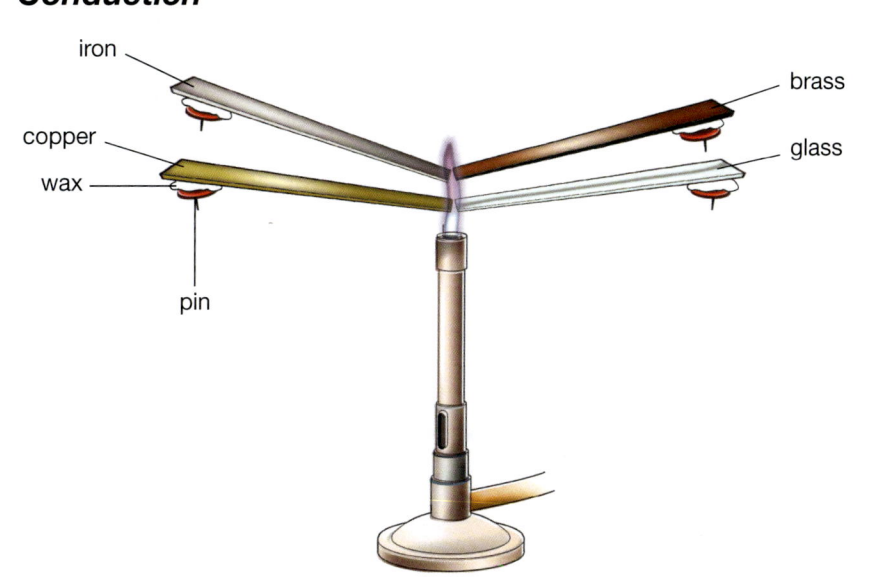

1. Select four strips of different material of equal length.

2. Attach a drawing pin to the end of each strip, using petroleum jelly or candle wax.

3. Heat the other ends of the strips equally.

4. The strip that allows the pin to drop first is the best conductor, and so on.

5. Glass is a bad conductor, so the drawing pin will stay attached.

Notes
IGCSE Physics Summarised

❑ **Convection**
1. Fill the container (see diagram) with cold water.

2. Carefully drop a few crystals of potassium manganate (VII) into the container.

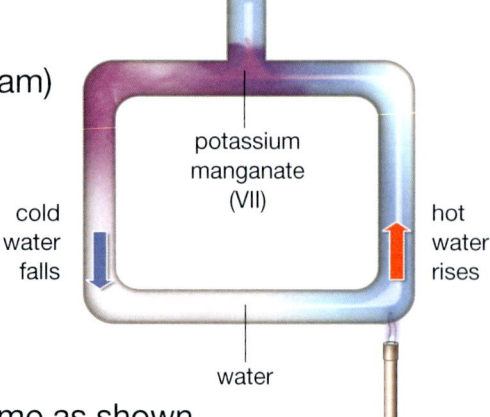

3. Heat gently with a small flame as shown.

4. The purplish pink colour moves in a circular path until all the water becomes coloured. Hot water rises and cold water falls.

○ **Radiation**
1. The diagram below shows two aluminium plates placed an equal distance from an electric heater. One of the plates is painted matt black and the other has a shiny polished surface. A cork is attached to the back of each plate with wax.

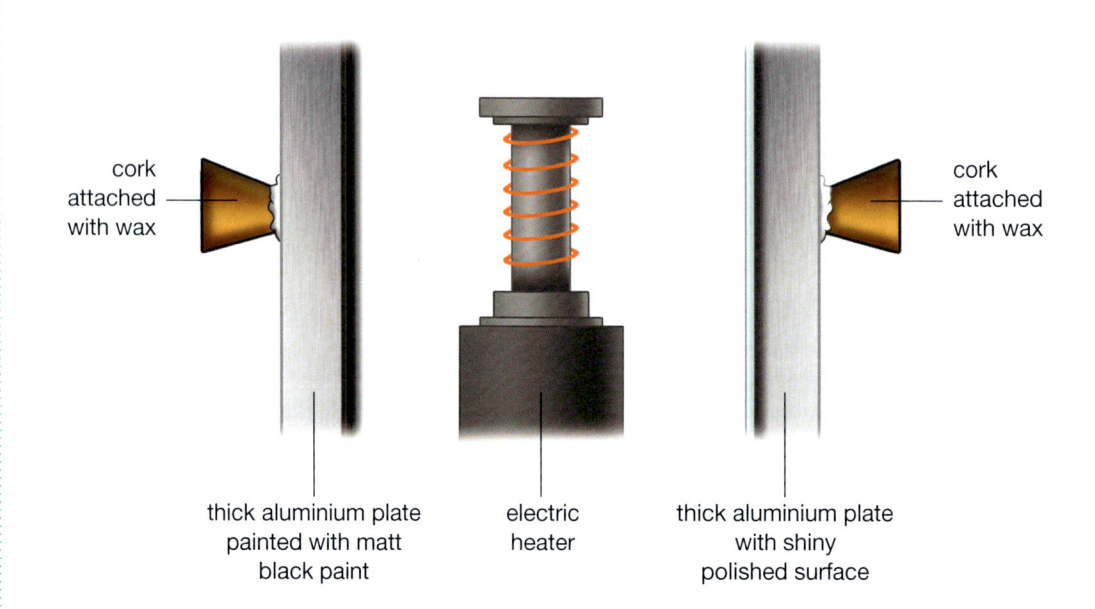

2. The cork attached to the matt black plate will drop off first, indicating that the matt black surface is a better **absorber** of thermal radiation than the shiny polished surface. Shiny surfaces are good reflectors of thermal radiation.

Unit 2 Thermal physics

○ The cube shown below is known as Leslie's cube. It has one shiny white surface, one dull white surface, one shiny black surface and one dull black surface.

rubber stopper

1. Fill the cube with boiling water and seal it with a stopper. All four sides of the cube are at the same temperature, because they are all in contact with the boiling water.

2. Place four thermometers or infra-red detectors a small distance from each of the surfaces. These will detect how much heat is radiated from each surface. Make sure the detectors are the **same distance** from each surface to ensure the experiment is fair.

3. The detector or thermometer at the dull black side will have the highest reading, followed by shiny black, dull white and shiny white. Therefore the dull black surface is the best emitter (i.e. the best surface for radiating heat), and the shiny white surface is the worst emitter.

❏ The table below gives a summary of the behaviour of surfaces.

	Best surface	**Worst surface**
Radiant heat emitters	dull black	shiny white
Radiant heat reflectors	shiny white	dull black
Radiant heat absorbers	dull black	shiny white

Consequences of energy transfer

❑ Some everyday examples of heat energy transfer are given below.

- White clothes are often worn in warm weather because they reflect infra-red radiation better.
- Highly polished teapots are not good radiators and so keep their contents warmer for longer than black teapots.
- We feel the heat radiated by the Sun, electric fires and electric lamps when our skin absorbs the radiation.
- White buildings keep cooler in warm weather than dark ones because they reflect the radiation from the Sun.
- If you heat a metal saucepan of water on top of the cooker, the heat transfers to the water by conduction. You can see the water moving as convection currents begin to transfer the heat through the water.
- The handle of a saucepan is often made of a poor conductor of heat such as wood or plastic so you don't burn your hand when lifting the pan.
- A cup of hot liquid will stay warmer in a polystyrene cup than in a metal one. In fact the process of cooling of a liquid is quite complex, but the biggest difference here is that the polystyrene is a poor conductor of heat and the metal is a good conductor so it is better at transferring the heat to the surroundings.

> ### Top Tip
>
> Any hot object will cool down more quickly if the temperature **difference** between the object and its surroundings is greater. For example, a cup of hot chocolate will cool down more quickly if the surrounding room is much colder than the drink. A smaller difference in temperature will result in a longer cooling time.

Reducing heat loss from the home

❏ When a house is heated in cold weather, there is heat transfer from the inside of the house to the surroundings. The diagram shows typical values for the proportions of the total heat lost through different parts of a house. (Of course the energy is not really 'lost'; it escapes from the house into the environment.) Heat loss at any given time depends on the difference in temperature between the inside and outside of the house. As the temperature difference increases, the rate of heat loss increases.

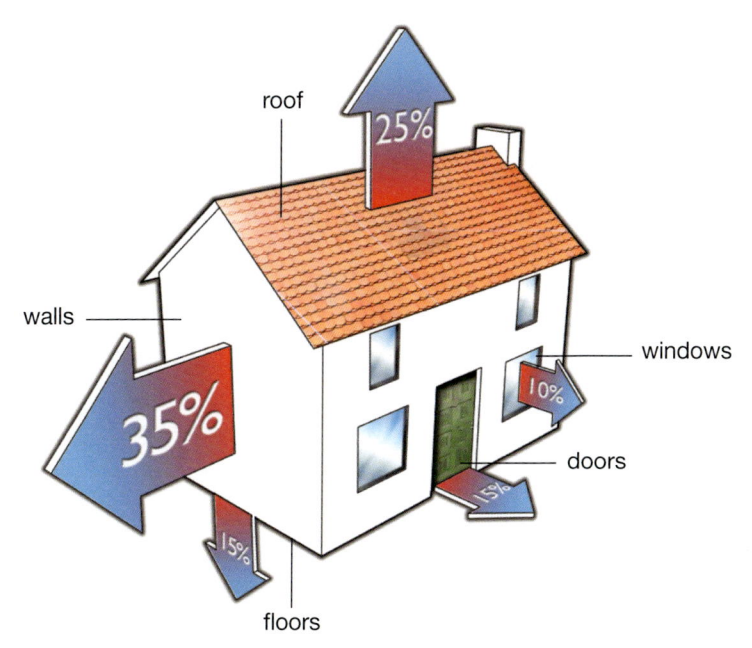

- roof 25%
- walls 35%
- windows 10%
- doors 15%
- floors 15%

❏ **Loft insulation** – Fibreglass reduces heat loss by conduction as it is a good insulator. Fibreglass also prevents heat loss by convection currents as the fibres trap the air and stop it rising.

❏ **Double glazing** – Air trapped between the sheets of glass reduces convection and conduction. Radiation will pass through unless the glass has a special reflective coating.

❏ **Floor insulation** – Carpets stop heat loss by conduction as the carpet fibres trap air. Air is a very good insulator.

❏ **Wall insulation** – Cavity walls (two layers of bricks with a gap between) can be filled with foam to prevent convection currents in the cavity as well as conduction through the walls.

The domestic radiator

❏ A radiator is really misnamed. It radiates some heat so, if you stand close enough, you can feel the infra-red radiation being emitted by the surface of the radiator.

However, most of the heat is taken away by the hot air that rises from the radiator.

Colder air from the room flows in to replace this hot air, and a **convection** current is formed as shown below.

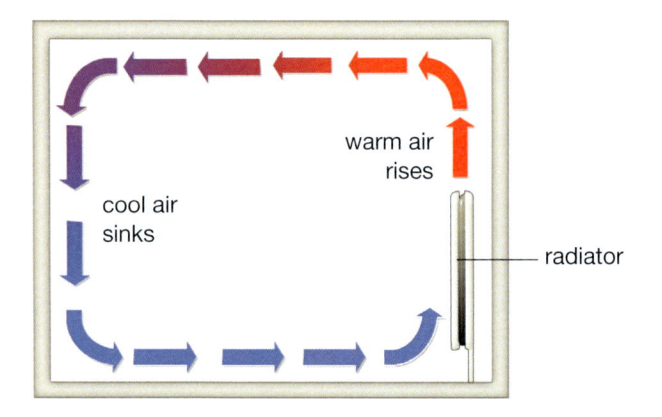

The vacuum flask

❏ A vacuum flask reduces conduction, convection and radiation. When a flask contains a hot liquid, the vacuum between the silvered surfaces (see the diagram) stops energy transfer by conduction and convection.

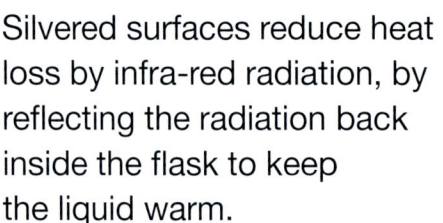

Silvered surfaces reduce heat loss by infra-red radiation, by reflecting the radiation back inside the flask to keep the liquid warm.

The cork at the top and the insulated supports at the bottom reduce heat loss by conduction. The cork at the top also reduces heat loss by convection and evaporation.

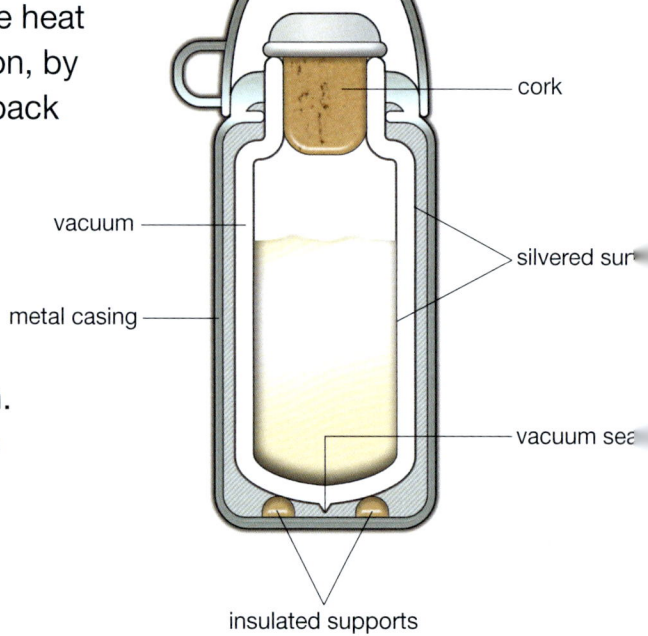

Unit 3 Properties of waves, including light and sound

Section 3.1 General wave properties

❑ **Waves** transfer energy from one place to another. Consequently, waves can be used to carry signals from one place to another.

❑ Waves are produced by vibrations.

❑ Waves have repeating patterns.

❑ **Wave terms**:

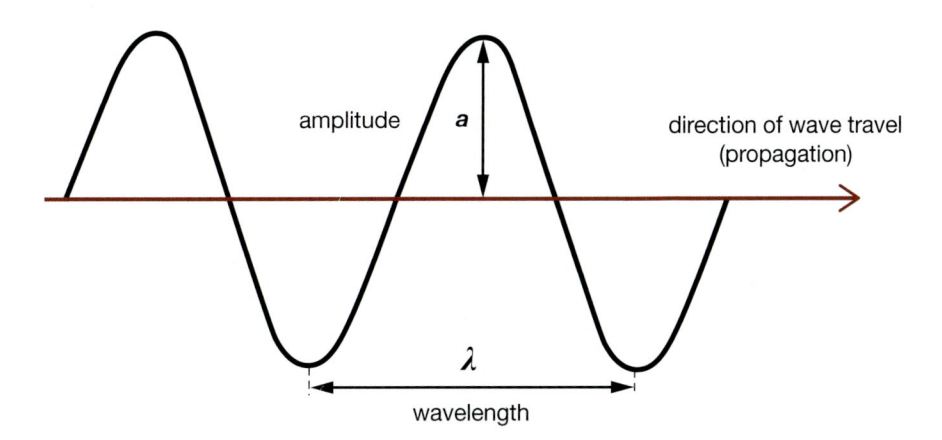

[Diagram of a sine wave showing amplitude a, wavelength λ, and direction of wave travel (propagation)]

- **amplitude** a – the maximum height of the wave from the **central equilibrium position**
- **wavelength** λ – the distance from any point on one wave to the same point on the next (adjacent) wave.

❑ Wave frequency is defined as the **number of wavelengths** produced or passing a specific point **per second**; it has the symbol f and its unit is the **hertz** (Hz).

> **Note**
>
> The larger the amplitude, the greater the energy a wave has.

- The **period** of a wave is the time taken for one wavelength to pass a specific point; it has the symbol **T** and its unit is the **second** (s).

$$T = \frac{1}{f} \implies f = \frac{1}{T}$$

- The speed of a wave **v** is the distance travelled by a particular point in a wave in one second.

- The wave formula is:

$$v = f\lambda$$

v = speed (m/s)
f = frequency (Hz)
λ = wavelength (m)

Types of waves

- There are two principal types of wave: longitudinal waves and transverse waves.

Longitudinal waves

- This type of wave can be shown by pushing and pulling a spring (slinky).

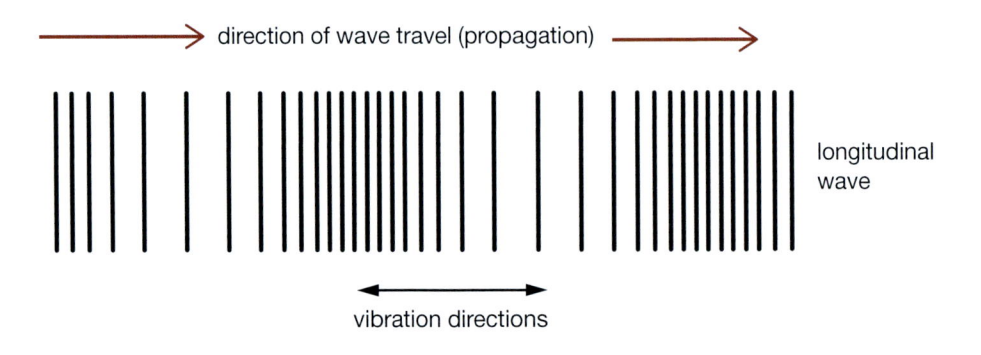

direction of wave travel (propagation) →

longitudinal wave

vibration directions

- In a longitudinal wave the **oscillations** or **vibrations** are **parallel** to the direction the wave energy is travelling (propagation direction).
- Longitudinal waves need a medium to travel through, and as they pass through the medium the particles oscillate backwards and forwards.
- **Sound** is an example of a longitudinal wave.

Transverse waves

❑ This type of wave can be shown by moving a spring (slinky) or a rope from side to side, or by making ripples in water.

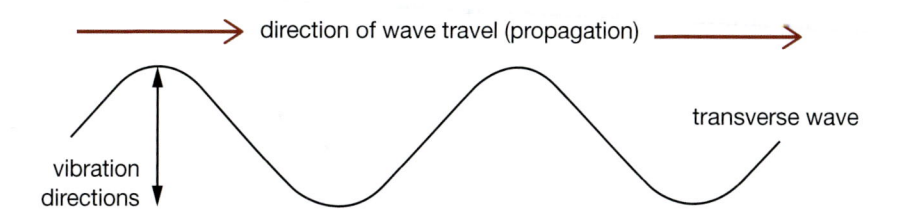

- In a transverse wave the oscillations or vibrations are at **right angles (perpendicular)** to the direction the wave energy is travelling (propagation direction).
- Transverse waves do not need a medium to travel through.
- **Light**, radio and other electromagnetic waves are transverse waves.

Reflection and refraction

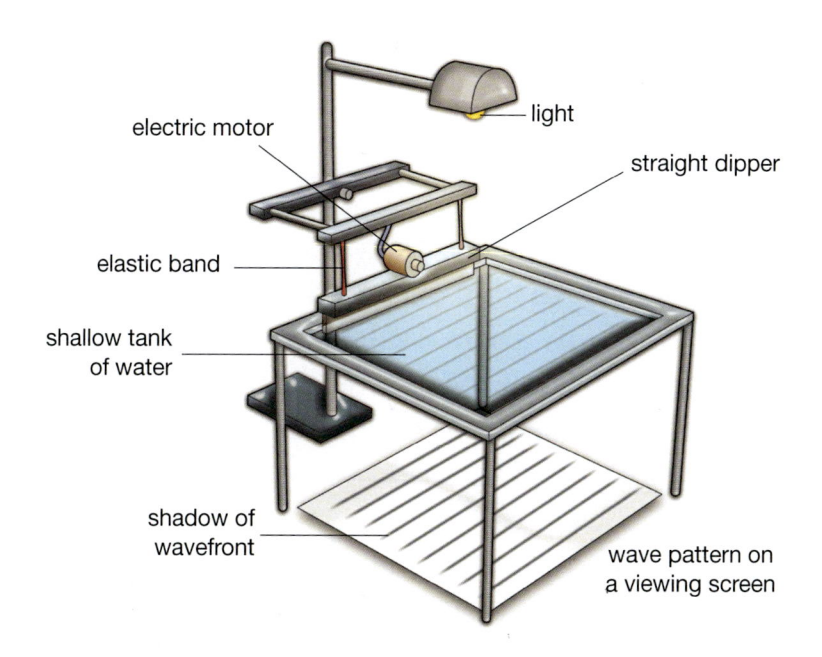

❑ The ripple tank, as shown in the diagram above, is used to produce waves. It can be used to study wave effects in different situations. The **wavefronts** can generally be thought of as **continuous lines** perpendicular to the direction of propagation. This is rather like viewing sea waves from the top of a cliff.

Reflection off a plane surface

❏ A **straight dipper** can be used to create **plane** waves.

❏ If the waves strike a plane barrier at 45° they are reflected as shown below.

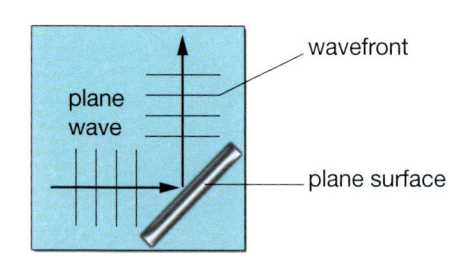

❏ The angle of incidence and the angle of reflection from a plane surface are **equal**.

❏ With reflection, **speed**, **frequency** and **wavelength** remain the **same**. Only the direction changes.

Refraction due to a change in speed

❏ A straight dipper can also be used demonstrate **refraction**. By placing a sheet of clear glass in the ripple tank the water above the glass will be shallower compared to the rest of the ripple tank.

❏ Keeping the frequency of the waves constant, it can be shown that when a wave moves from one depth into another, it will either **speed up** or **slow down**. Water waves travel faster in deep water, and slower in shallow water. Notice (in the diagram at the top of the opposite page) that the wavefronts are closer together in the slower region.

❏ When a wave travels from deep water to shallow water:
- wavelength decreases
- speed decreases
- frequency stays the same.

❏ When a wave moves from one depth to another at an **angle**, it will **change direction**. We say it has been **refracted**.

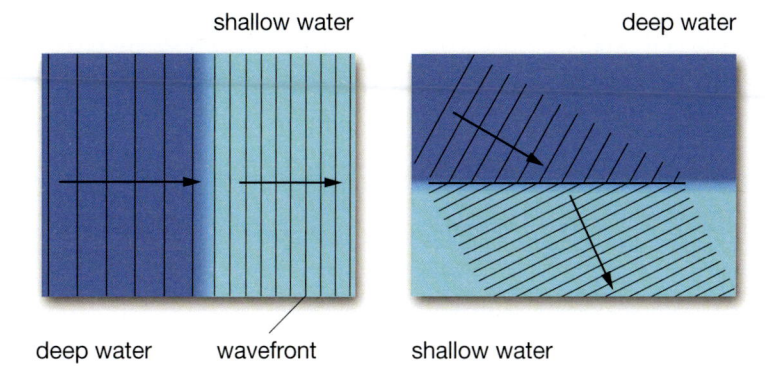

Diffraction

❏ Diffraction occurs when a wave spreads out as it passes through a gap.

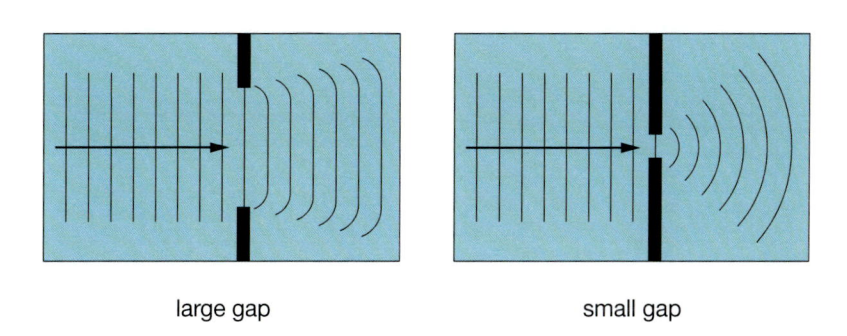

○ Waves diffract when they pass an edge. The wavefront becomes bent. The longer the wavelength, the more the wavefront is bent.

○ If a gap is about the same size as the wavelength there is maximum diffraction and parallel wavefronts emerge as circular wavefronts (see 'small gap' diagram above).

○ **Sound waves** have long wavelengths and they diffract through large angles, hence we can hear sound around corners. **Light waves** have very short wavelengths and so **they diffract through negligibly small angles**, hence we cannot see or be seen around corners.

○ TV and FM radio reception is sometimes poor for people who live in hilly regions. The waves carrying the signal have a wavelength much smaller than the hill, and as they pass over the hill there is little diffraction. The people who live in the valley may not receive the signal. Long wavelength (low frequency) radio waves will be diffracted more because the wavelength is comparable to the hill size, and they may be received as shown below.

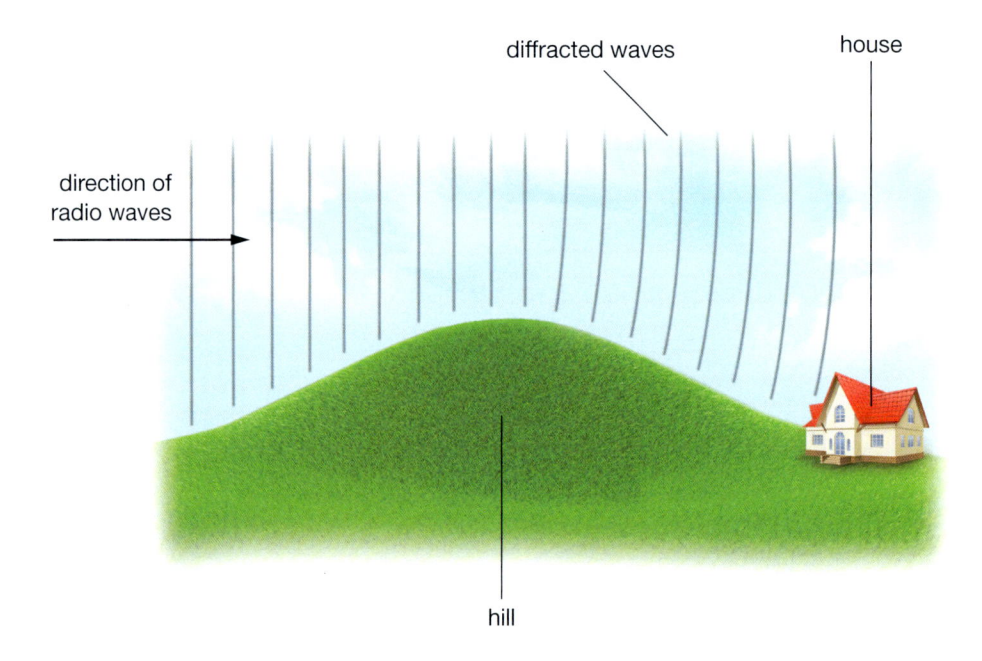

diffracted waves

house

direction of radio waves

hill

Note

During diffraction, **speed**, **frequency** and **wavelength** stay the same.

Top Tip

When sketching diffraction diagrams, make sure the wavelength is the **same** either side of the gap. The emergent waves should be **curved** at the edges if the gap is **large** and **circular** if the gap is **small**, as shown on page 129.

Unit 3 Properties of waves

Section 3.2 Light

Reflection of light

- Properties of light:
  - it travels as **transverse** waves
  - it transfers energy
  - it can travel in a vacuum
  - it travels at a speed of 3.0×10^8 m/s in air or a vacuum.

- A light ray is a narrow beam of light that travels in a straight line. A light ray that reflects from a surface such as a **plane mirror** obeys the **law of reflection**.

- The angle of incidence is equal to the angle of reflection, where both angles are measured to the **normal**.

- The normal is a construction line at **90°** to the mirror, at the point where the light ray meets the mirror. It allows the angles shown in the diagram to be measured.

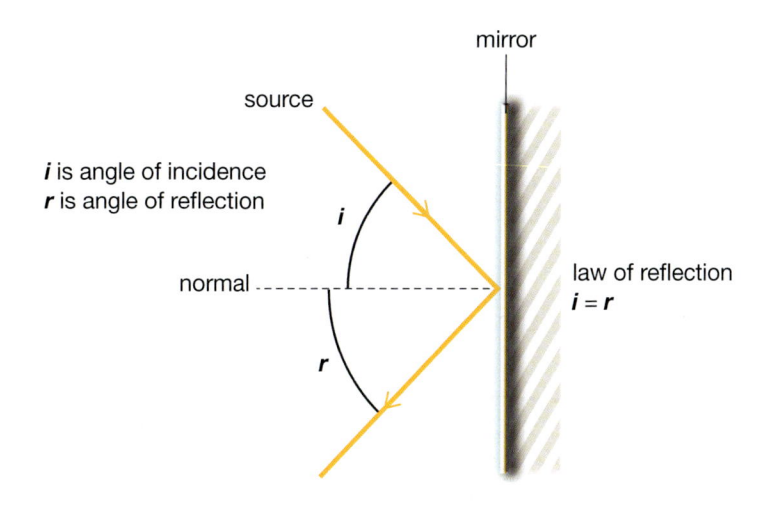

i is angle of incidence
r is angle of reflection

law of reflection
$i = r$

angle of incidence (*i*) = angle of reflection (*r*)

> **Note**
>
> During reflection, **speed, frequency** and **wavelength** do not change; only **direction** changes.

❏ A plane mirror forms an **image** of an object, which has these properties:
- upright but **laterally inverted** i.e. the image is reversed left to right
- the same size as the object
- the same distance behind the mirror as the object is in front.

❏ Plane mirrors are used in periscopes, security mirrors and dressing table mirrors.

○ **Finding the position of the image using a plane mirror**

1. Draw any two incident rays from the object to the mirror.

2. Draw in the reflected rays from these incident rays, making sure **angle of incidence = angle of reflection**.

3. Continue these reflected rays back straight behind the mirror using a dotted line to the point where they appear to come from.

4. The image is formed where the rays appear to come from behind the mirror. This is a **virtual image**. A virtual image cannot be formed on a screen. If you placed a screen at position 4 in the diagram below nothing would show.

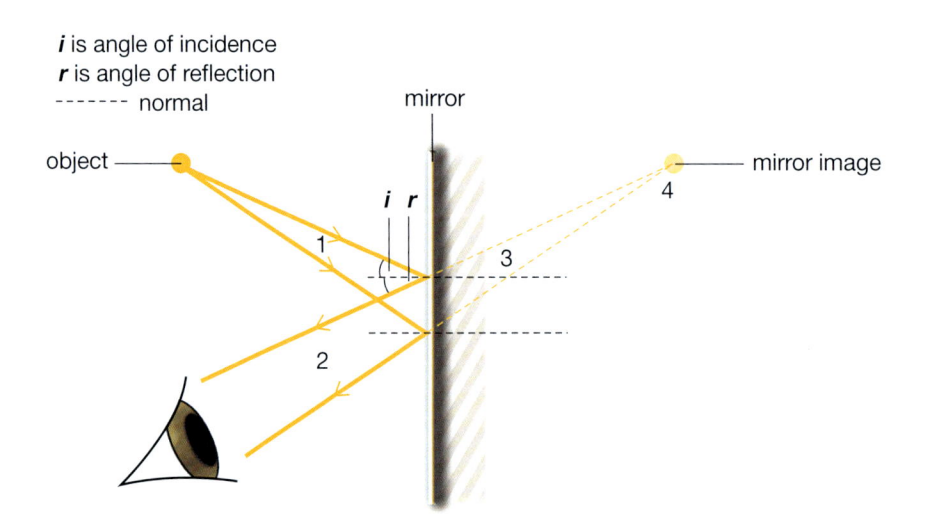

i is angle of incidence
r is angle of reflection
------- normal

○ ***Example***
(a) An object is placed in front of a mirror. Construct a ray diagram showing where the position of the image will appear.
(b) Describe two properties of the image.

Unit 3 Properties of waves

Notes

(a) 1. Draw a ray A from the top of the object to the mirror. Mark in the normal and measure, using a protractor, the angle of incidence *i*.

2. Construct the reflected ray such that the angle of incidence *i* is equal to the angle of reflection *r*.

3. Draw a second ray B in the same way.

4. Use dotted lines to show where the reflected rays appear to come from. This will be the top of the image.

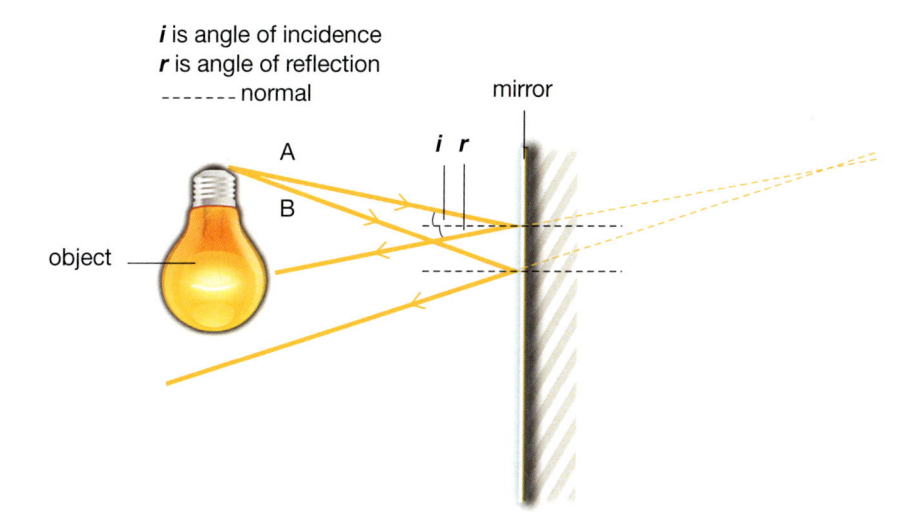

i is angle of incidence
r is angle of reflection
------- normal

5. Repeat with two rays, C and D, from the bottom of the object.

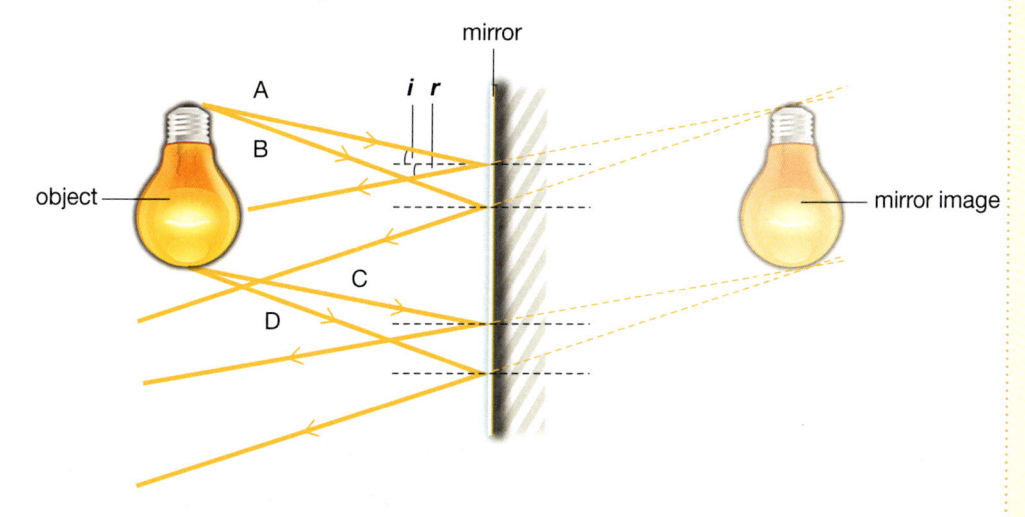

(b) - The image is upright (the same way up as the object).
- It is the same size as the object.

133

Refraction of light

❏ Light rays can **change direction** when passing from one material into another because the rays move at **different speeds** in the two materials.

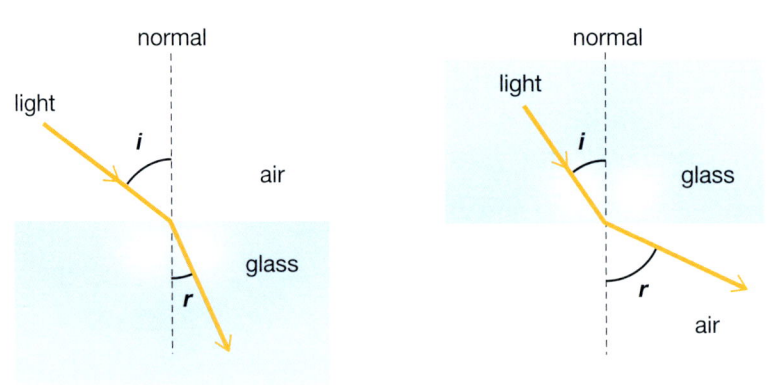

❏ Notice that the angle of incidence *i* and the angle of refraction *r* always lie between the light beam and the normal.

❏ A material in which light travels slowly is said to be **optically dense**.

❏ When travelling from an optically **less dense** material to a **more dense** material (such as air to glass), light **bends towards** the **normal** (diagram on the left above).

❏ When travelling from an optically **more dense** material to a **less dense** material (such as glass to air), light bends away from the **normal** (diagram on the right above).

❏ When light enters a different material:
   - the frequency stays the same
   - the wavelength changes
   - the speed changes.

> **Top Tip**
>
> When a light ray travels **along the normal** from air into glass, **it passes straight through** (undeviated), but it slows down.

Refractive index

○ The **refractive index** n of a material indicates how strongly the material changes the direction of light. It is one of the few quantities that **does not have a unit**; it is just a number.

○ Refractive index, the speed of light in a vacuum (or air) and the speed of light in a medium (e.g. a material like glass, water or Perspex) are related by the formula:

$$n = \frac{\text{speed of light in air (or vacuum)}}{\text{speed of light in material}}$$

(Speed of light is given the symbol c.)

so $n = \dfrac{c_v}{c_m}$

○ Refractive index, the angle of incidence and the angle of refraction are related by the formula:

$$\boxed{n = \frac{\sin i}{\sin r}}$$

n = refractive index (no units)
i = angle of incidence (°)
r = angle of refraction (°)

This relationship is **Snell's Law**.

> **Note**
>
> Refractive index is **greater** than 1 ($n>1$) when light goes from a **less dense** to a **more dense** medium.

Example 1

Light travels from air into a glass block of refractive index 1.4. The angle of refraction in the glass is 35°. Calculate the angle of incidence in air.

Step 1 List all the information in symbol form and change into appropriate and consistent SI units if required.

$n = 1.4$
$r = 35°$
$i = ?$

Step 2 Use and rearrange the correct formula.

$$n = \frac{\sin i}{\sin r} \quad \Rightarrow \quad \sin i = n \times \sin r$$

$$i = \sin^{-1}(n \times \sin r)$$

(The term $\sin^{-1}(x)$ means the angle whose sine value is x.)

Step 3 Put the numbers into the formula and calculate the answer.

$$i = \sin^{-1}(1.4 \times \sin 35°) = 53°$$

ALWAYS REMEMBER TO STATE THE UNIT FOR CALCULATED QUANTITIES.

Top Tip

When a light ray travels from a **more dense** material (glass) to a **less dense** material (air) the **refractive index must be less than 1** ($n < 1$).

Make sure you know which way the light is travelling. If you are given $n > 1$ for light travelling from air to glass, but in the question the light is travelling from glass to air, you must first calculate the correct refractive index by finding the **reciprocal**.

If $n_{\text{air to glass}} = 1.43$ then $n_{\text{glass to air}} = 1/1.43 = 0.7$

Once you have the correct refractive index n, the formula above can be used as normal.

Example 2

The speed of light in a vacuum is 3.0×10^8 m/s. Calculate the speed of light in glass of refractive index 1.6.

Step 1 List all the information in symbol form and change into appropriate and consistent SI units if required.

$n = 1.6$
c_v = speed in vacuum = 3.0×10^8 m/s
c_m = speed in material = ?

Step 2 Use and rearrange the correct formula.

$$n = \frac{\text{speed of light in vacuum}}{\text{speed of light in material}}$$

$$n = \frac{c_v}{c_m} \implies c_m = \frac{c_v}{n}$$

Step 3 Calculate the answer by putting the numbers into the formula.

$$c_m = \frac{3.0 \times 10^8}{1.6} = 1.9 \times 10^8 \text{ m/s}$$

ALWAYS REMEMBER TO STATE THE UNIT FOR CALCULATED QUANTITIES.

Measuring angles of incidence and refraction

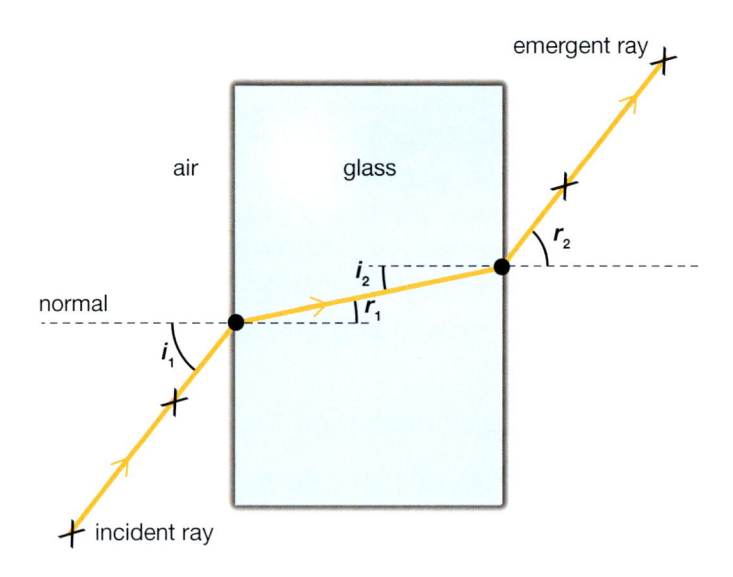

❑ The following experiment describes how to measure the angle of incidence and angle of refraction.

1. Place a transparent block in the middle of a plain sheet of paper; trace around the block in pencil.
2. Position a raybox so that the light from it strikes the glass block at an angle. (**Remember**: If the light strikes the glass boundary at 90°, it will pass straight through undeviated.)
3. Mark the positions where the light meets the glass boundary and where it leaves the glass boundary with dots (see the diagram).
4. Mark two crosses (or place optical pins) on the paper along the incident ray and the emergent ray approximately 5 cm apart.
5. Remove the glass block and switch off the raybox.
6. Using a ruler, complete the lines between the dots and the crosses.
7. Draw in the normal at **90°** to where the light strikes the glass boundary.
8. Draw the second normal where the light leaves the glass boundary, again at **90°**.
9. Using a protractor, measure the angles of incidence and refraction as shown on the diagram above.

Unit 3 Properties of waves

Total internal reflection

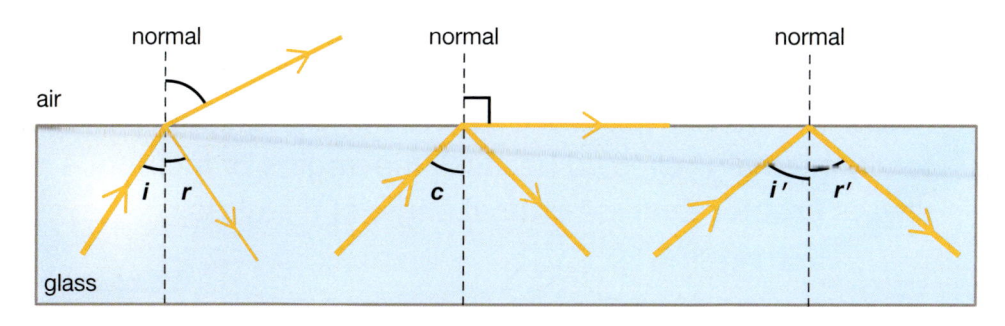

| For small angles of incidence, the ray splits into a refracted ray and a weak internally reflected ray. | At the critical angle c, most of the light is refracted at 90° along the glass surface; the internally reflected ray has become a little stronger. | For angles of incidence greater than the critical angle, **total internal reflection** occurs. No light is refracted. |

- ❑ **Total internal reflection** occurs when **light is travelling from a more dense to a less dense material** and the angle of incidence is greater than the **critical angle** of the material.

- ❑ The **critical angle** is defined as the angle in the denser material above which total internal reflection occurs.

- ❑ Each material has its own critical angle. For example, the critical angle of glass is 42° and for water it is 49°.

- ❑ Consequently, light incident at angles greater than 42° for glass and greater than 49° for water will be completely reflected and none of it will be refracted – total internal reflection (see page 140).

- ❑ The inside surface of diamond, water or glass can act like a mirror depending on the angle at which light strikes it.

- ○ The critical angle c is related to the refractive index n by the formula:

$$n = \frac{1}{\sin c}$$

n = refractive index (no units)
c = critical angle (°)

Uses of total internal reflection

○ Total internal reflection is used in **fibre optic cables**. A fibre optic cable is made up of a bundle of very thin glass fibres. The light travels along the fibre by being constantly totally internally reflected because the **angles of incidence** are always **greater** than the **critical angle** of the glass.

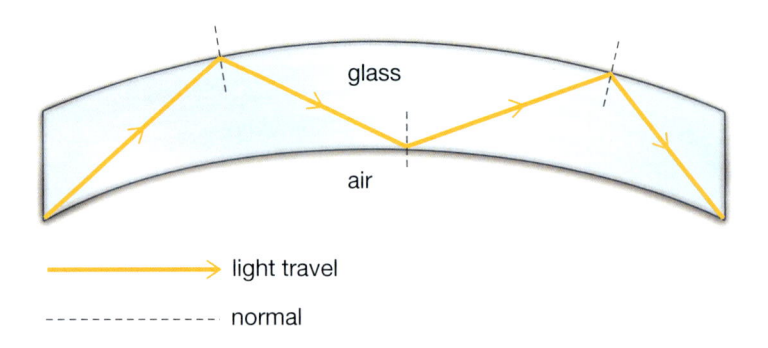

○ Fibre optic cables are used for TV transmission, telephone cables, internet communications and in medical devices such as endoscopes. An endoscope consists of a long, thin, flexible tube which has a light and a video camera. Images of the inside of the patient's body can be seen on a screen.

❏ The inside of a glass prism can be used as a mirror. Total internal reflection takes place on the **longest** face of the prism, as shown in the diagram, and occurs because the angle of incidence (45°) is greater than the critical angle (42°). This arrangement is used in **periscopes**.

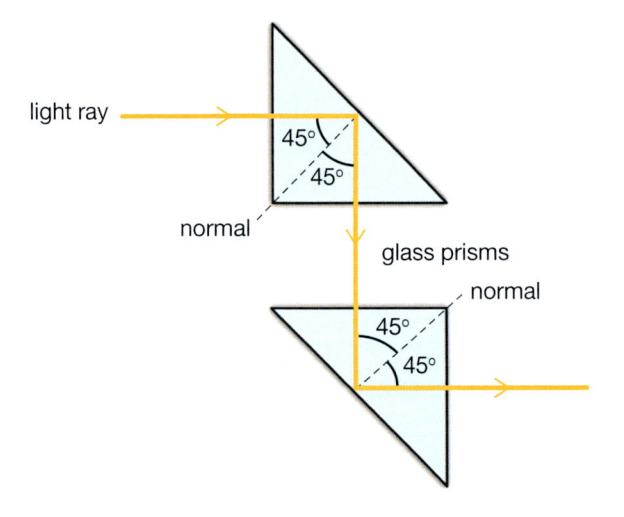

❏ Pairs of internally reflecting prisms are also used in **binoculars**.

Thin converging lens

❏ Lenses bend light to form images. The **converging** (convex) lens is shown below:

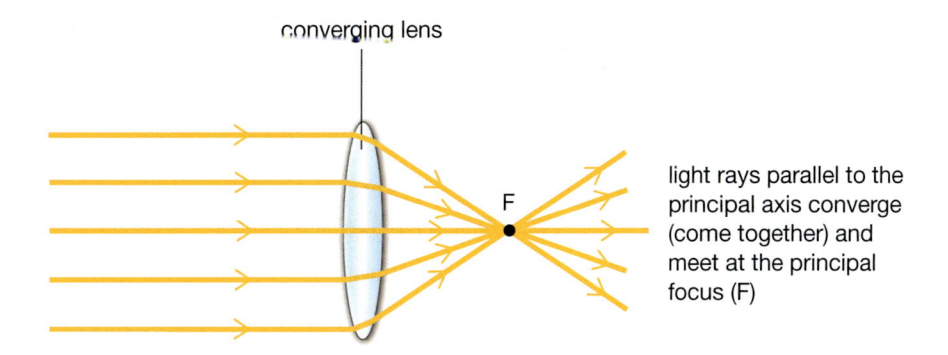

converging lens

light rays parallel to the principal axis converge (come together) and meet at the principal focus (F)

❏ A converging lens converges a light beam (**brings rays together**) to produce a focused image.

❏ The image formed by rays converging is a **real image**, which means it can be formed on a screen (see top diagram page 142).

❏ When a beam of light parallel to the principal axis passes through a converging lens, it is refracted so that it converges at a point known as the **focal point** or **principal focus**. There are two principal foci, one on each side of the lens.

❏ The **focal length f** of a lens is the distance between the **centre** of the lens and the **principal focus F**.

❏ The points **2F** are both a distance **2f** from the centre of the lens.

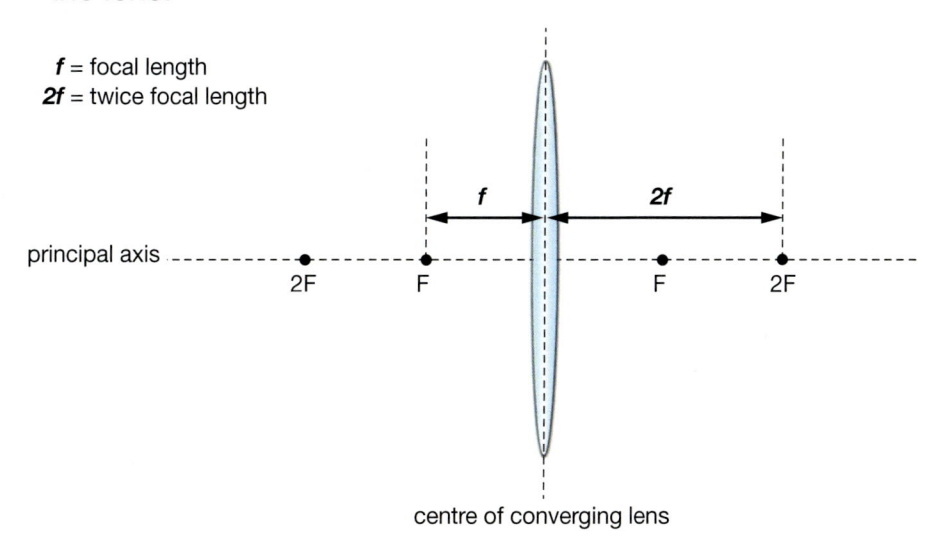

f = focal length
2f = twice focal length

principal axis
2F   F f 2f F   2F

centre of converging lens

Drawing ray diagrams

❏ For a converging lens, to find the position of an image:

1. Draw a ray from the top of the object through the centre of the lens, which does not change direction.

2. Draw a ray from the top of the object parallel to the principal axis until it reaches the central plane of the lens. The ray then passes straight through the principal focus on the other side of the lens. An image is formed where the rays meet.

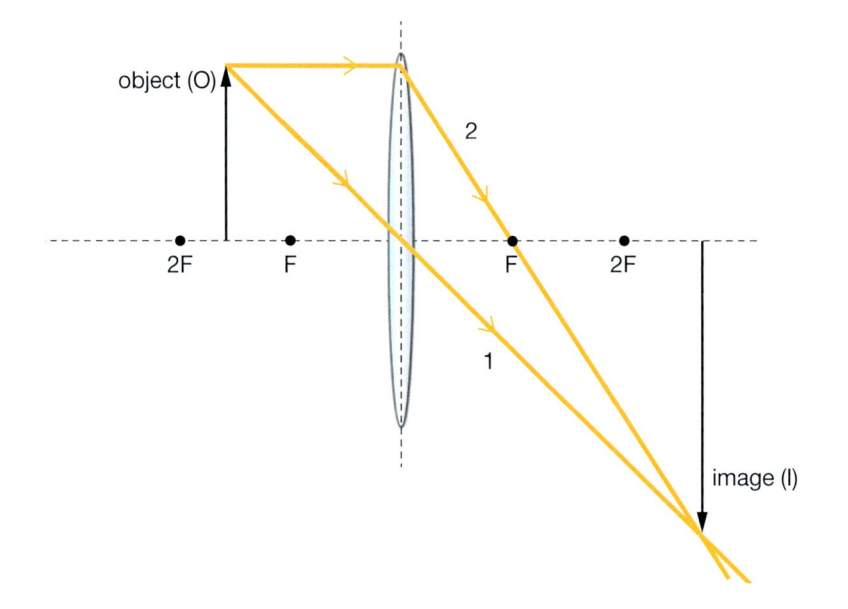

○ A converging lens can be used as a **magnifying glass** when an object is placed between **F** and the lens.

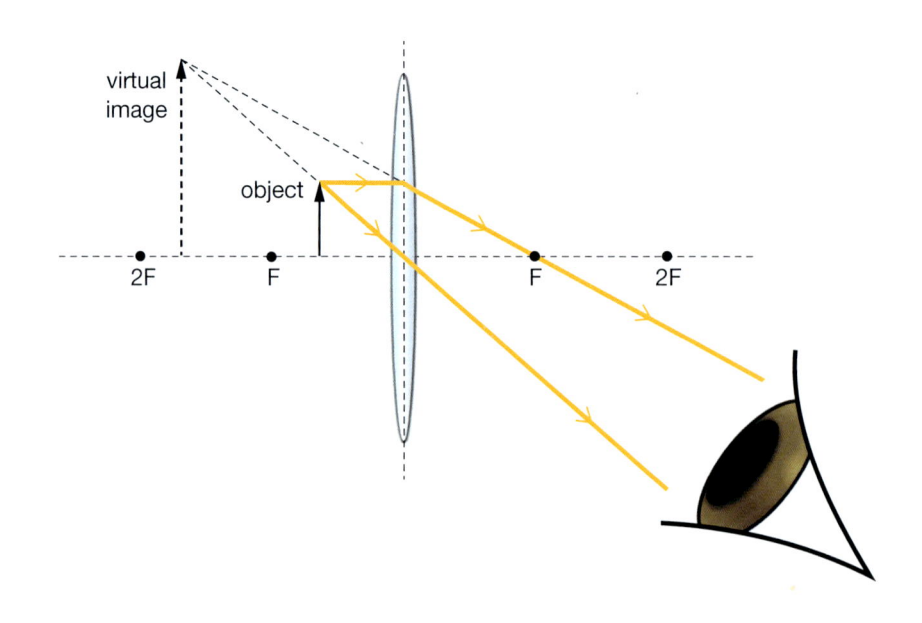

○ In this case the rays do not meet when following the ray diagram rules explained at the top of page 142. When the lines are extended backwards (extrapolated), they appear to meet as shown by the dotted lines in the diagram at the bottom of that page. This is known as a **virtual image**, which can be seen by positioning a human eye as shown.

> **Note**
>
> A **real image** forms where the light converges and it can be formed on a screen.
>
> A **virtual image** is where all the light appears to come from and cannot be formed on a screen.

❏ Images can also be described as:
- upright (same way up) or inverted (upside down)
- magnified (larger) or diminished (smaller).

> **Top Tip**
>
> Sometimes you may be asked to draw a third ray when drawing a ray diagram. The rules are the same; the top ray goes parallel to the principal axis and then through F. The middle ray goes through the centre of the lens without changing direction. The bottom ray goes through F and then goes parallel to the principal axis as shown below.

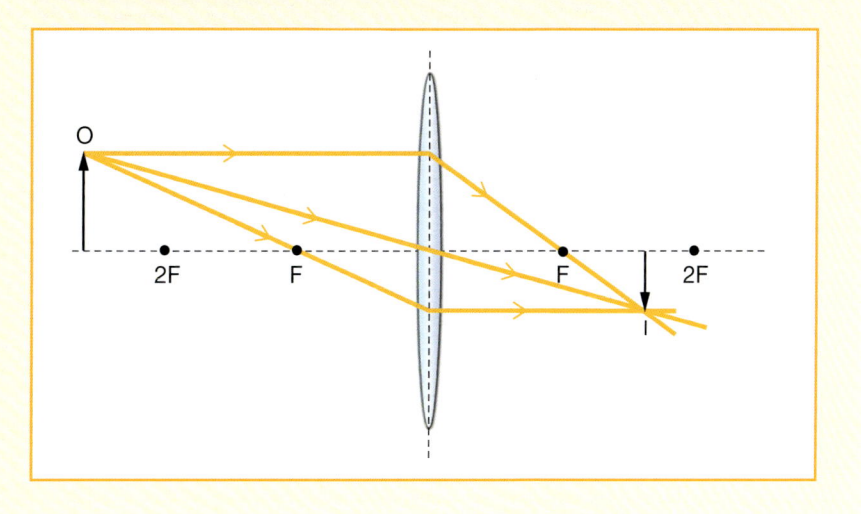

Formation of images by a converging lens

❏ **Object:** Further than **2f** from lens **Image:** **Uses:**
- between **f** and camera
  2f from lens
- inverted
- diminished
- real

❏ **Object:** Between **f** and **2f** from lens **Image:** **Uses:**
- more than **2f** projector
  from lens
- inverted
- magnified
- real

○ **Object:** Less than **f** from lens **Image:** **Uses:**
- rays do not magnifying
  converge glass
- virtual
- upright
- magnified

Dispersion of light

- Light of different colours has different wavelengths.

- **Red** light has a longer wavelength than **green** light, which has a longer wavelength than **blue** light.

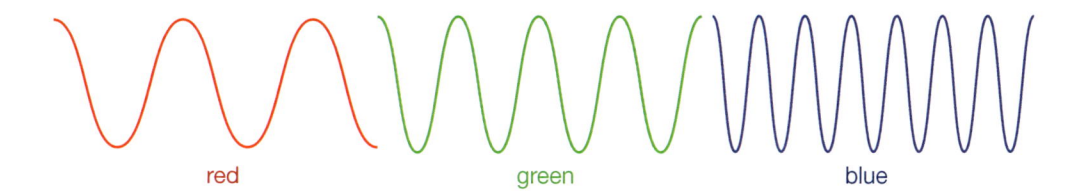

  red   green   blue

- Light of a single wavelength and therefore a single frequency, is known as **monochromatic** light. **Lasers** produce monochromatic light.

- **White** light is said to be made up of **seven** different colours. The colours travel at the same speed in space (vacuum) but slow down when they enter a denser medium like glass. The different wavelengths slow down by different amounts and so refract by different angles in a prism. This is known as dispersion and a **spectrum** is produced.

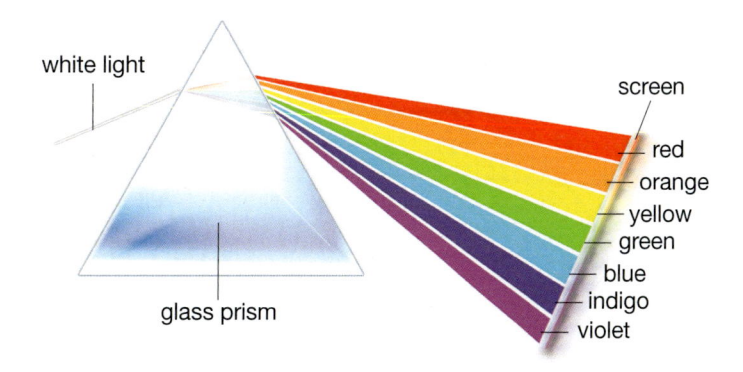

  white light → glass prism → screen: red, orange, yellow, green, blue, indigo, violet

- The colours of the **visible light spectrum** can be memorised using the mnemonic: **ROYGBIV**: Richard Of York Gave Battle In Vain.

> *Top Tip*
>
> Always describe white light **splitting up** into a **spectrum of light** and **not** 'colours of the rainbow'.

Section 3.3 Electromagnetic spectrum

❑ The electromagnetic (e.m.) spectrum is a family of **electromagnetic waves** that travel at the same high speed; they have different wavelengths and frequencies.

❑ Radio waves, microwaves, infra-red, visible light, ultraviolet, X-rays and gamma (γ) radiation are all parts of the electromagnetic spectrum.

❑ The different wavelengths of the different types of wave are shown in the diagram below. Note, for example, that infra-red radiation, responsible for the transfer of heat energy by radiation, has a slightly longer wavelength than visible light.

❑ Electromagnetic waves can travel through a vacuum.

❑ All electromagnetic waves are **transverse waves** and transfer energy but not matter.

○ The **speed** of **all** electromagnetic waves in a vacuum (or approximately in air) is 3.0×10^8 m/s.

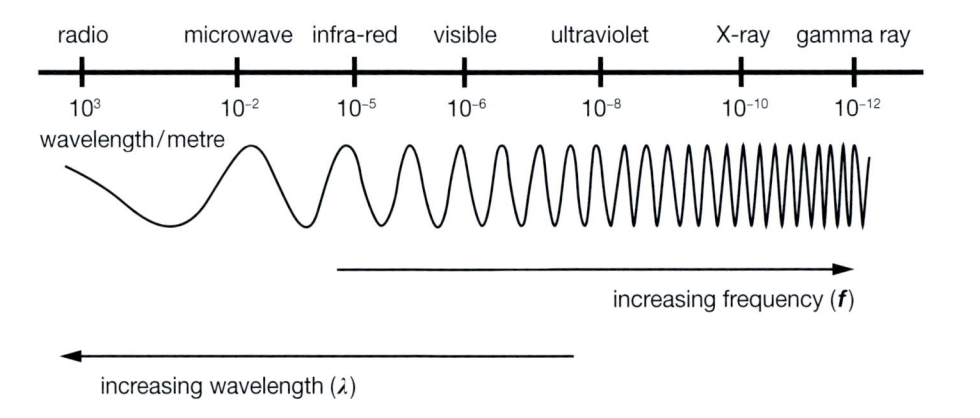

radio	microwave	infra-red	visible	ultraviolet	X-ray	gamma ray
10^3	10^{-2}	10^{-5}	10^{-6}	10^{-8}	10^{-10}	10^{-12}

wavelength/metre

increasing frequency (**f**)

increasing wavelength (λ)

❑ Exposure to large doses of any type of electromagnetic radiation can be harmful. Microwaves, of the frequency used in microwave ovens, cause internal heating of body tissues and may cause burning as may infra-red radiation.

Even small doses of short wavelength e.m. radiation can be harmful. Both X-rays and gamma rays damage living cells and can cause the death of living cells.

Summary of different electromagnetic waves

- ❏ *Gamma (γ-) rays*
    - very short wavelength
    - dangerous (γ-rays can kill living cells)
    - pass through skin and soft body tissue
    - come from radioactive substances such as uranium
    - used to kill cancer cells and sterilise hospital equipment

- ❏ *X-rays*
    - short wavelength
    - dangerous (X-rays can kill living cells)
    - pass through skin and soft body tissue but not bone or metal
    - used to photograph broken bones and in security systems

- ❏ *Ultraviolet rays*
    - wavelength a little shorter than visible light
    - causes tanning and can damage the skin if over-exposed
    - used to check for forged bank notes

- ❏ *Visible light*
    - mid-range wavelength
    - composed of a spectrum of colours
    - comes from the Sun and other luminous objects

- ❏ *Infra-red rays*
    - wavelength a little longer than visible light
    - come from the Sun and any other hot objects
    - used in TV remote controls and security alarms

- ❏ *Microwaves*
    - long wavelength
    - used in mobile phones, microwave ovens, satellite television and communication satellites

- ❏ *Radio waves*
    - very long wavelength
    - used for terrestrial television and radio broadcasts

Section 3.4 Sound

- Sound is caused by **vibrations**. It travels as a wave through a medium (which may be **solid**, **liquid** or **gas**) and transfers energy. Examples of how sound is produced include:
    - a hammer hitting a nail: the hammer and the nail vibrate, which in turn cause air to vibrate, creating a sound wave that travels to the ear
    - a door slamming: the door and its frame vibrate, causing air molecules to vibrate, creating a sound wave that travels to the ear
    - a cricket bat hitting a ball: the bat and ball vibrate, causing air molecules to vibrate, creating a sound wave that travels to the ear.

- Sound **cannot** travel through a **vacuum**. (There are no particles to vibrate.)

- The **speed of sound in air** is approximately **330 m/s**.

- Sound waves have a wide range of wavelengths and therefore frequencies. Our ears, however, do not respond to all frequencies. The normal **human hearing** frequency range is **20 Hz – 20 000 Hz**.

- Sound above 20 000 Hz is known as **ultrasound**. In medicine an ultrasound scan (sometimes called a sonogram) is a painless test that creates images of organs and structures inside your body using sound waves.

- Sound with a frequency lower than 20 Hz is called infrasound. Infrasound is used by some large animals for communication. For example, whales, giraffes and elephants can communicate over many miles using infrasound.

- Sound waves are **longitudinal waves**; their vibrations are **parallel** to the direction of propagation (see page 126).

Unit 3 Properties of waves   Notes

○ **Sound waves** are a series of **compressions** and **rarefactions**.

- Areas where there is high pressure (where the molecules are squashed together) are known as **compressions**.
- Areas where there is low pressure (where the molecules are further apart) are known as **rarefactions**.

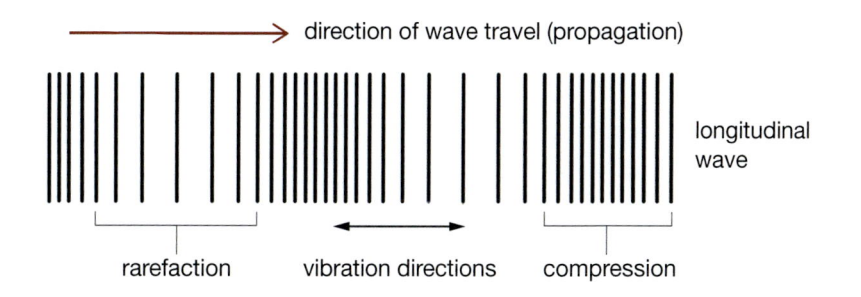

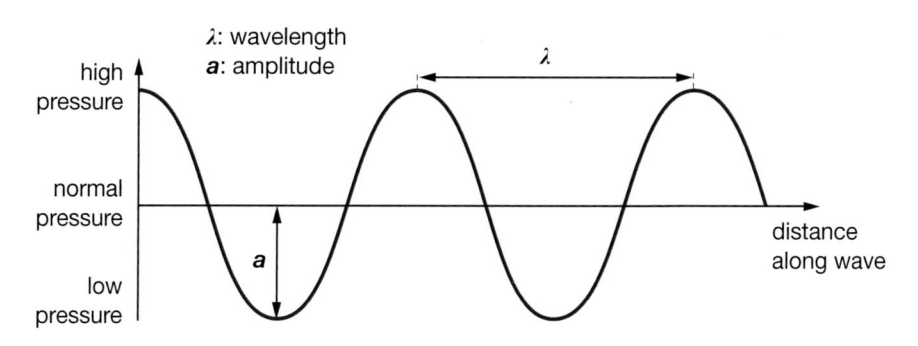

→ direction of wave travel (propagation)

longitudinal wave

rarefaction   vibration directions   compression

λ: wavelength
a: amplitude

high pressure
normal pressure
low pressure

distance along wave

Top Tip

Sometimes questions about sound can show what looks like a **transverse wave**; be careful as it is a **graphical** representation of **pressure** in a sound wave. Notice in the diagram above how the peaks of the graph correspond to the positions of compression in the longitudinal wave.

Speed of sound

❏ The following experiment describes how to determine the speed of sound in air.

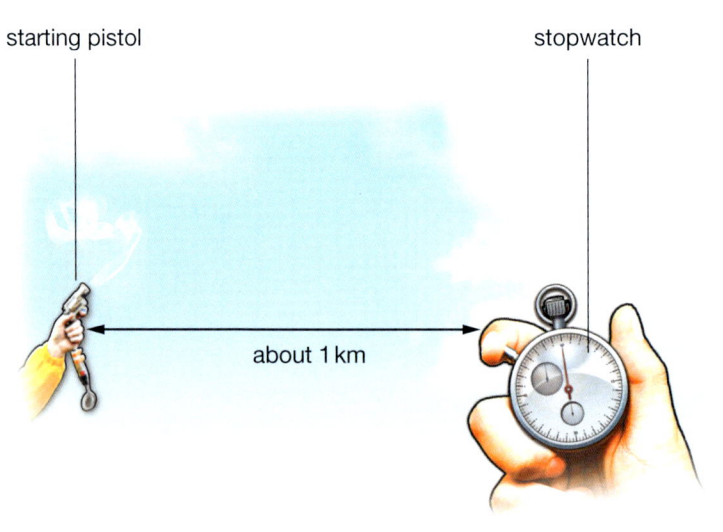

1. A student with a stopwatch stands **a long distance** from a teacher with a starting pistol.

2. The teacher fires the pistol in the air.

3. The student starts the stopwatch when he/she sees the puff of smoke and stops it when he/she hears the bang.

Calculations to be made
Use the formula to calculate the speed of sound:

$$\text{speed} = \frac{\text{distance}}{\text{time}} \qquad v = \frac{d}{t}$$

Improvements to the experiment
The experiment could be improved by:

1. repeating it to find an average value, and using different students

2. increasing the distance between the pistol and the stopwatch, so that the timekeeper's reaction time has less effect on the time

3. swapping the position of the pistol and the stopwatch to take account of any wind direction, because the direction of the wind will affect the speed of sound.

Unit 3 Properties of waves

> **Note**
>
> The speed of light (3.0×10^8 m/s) is very much greater than the speed of sound (330 m/s). This is why we see lightning first and then hear thunder a little later.

The speed of sound and echoes

- Sound is a longitudinal wave that can travel to a surface, be reflected and travel back again as an **echo**.

- If the distance between the source of sound and the reflective surface, and the time taken, are known then the speed the wave has travelled can be calculated using $v = d/t$.

- **Example**

  Sara stands 200 m from a large wall and claps together two large blocks of wood. She hears an echo. She claps regularly and times her claps so that they coincide with the echoes. A friend standing beside her with a stopwatch records the time taken for the sound of 10 claps to reach the wall and echo back again as 12.4 s. Calculate the speed of sound.

  Step 1 List all the information in symbol form and change into appropriate and consistent SI units if required.

  $d = 2 \times 200$ m (the distance to the wall and back)
  The time taken for sound to travel there and back for 10 claps is 12.4 s, so

  $t = 1.24$ s

  Step 2 Use the correct formula.

  $$v = \frac{d}{t}$$

  Step 3 Calculate the answer by putting the numbers into the formula.

  $$v = \frac{2 \times 200}{1.24} = 320 \text{ m/s} \text{ (to 2 significant figures)}$$

ALWAYS REMEMBER TO STATE UNITS OF CALCULATED QUANTITIES

- ❏ If the speed of sound in a medium and the time taken are known then the distance the wave has travelled can be calculated using **d = vt**.

- ❏ **Example**
  A fishing boat uses sonar (sound, navigation and ranging) to detect shoals of fish under the water. A ping of sound is sent from the surface, and is reflected back after it hits the shoal of fish. Sound travels at 1500 m/s in water and the time between the ping and its echo is 0.40 s. Calculate the depth of the water where the shoal is located.

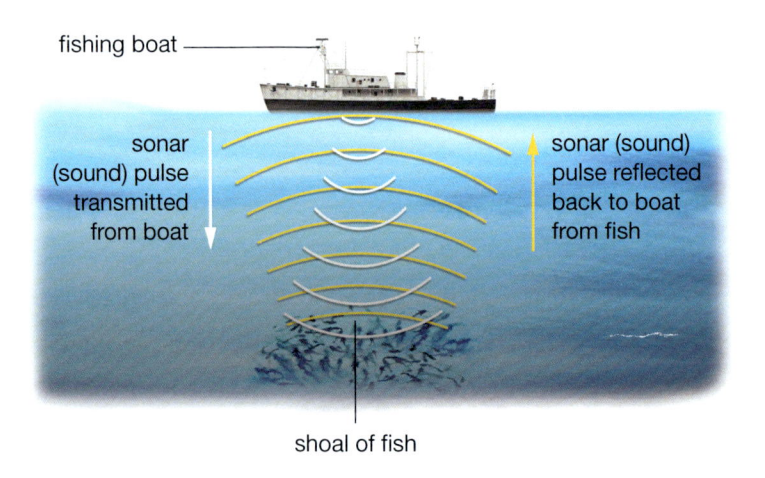

Step 1 List all the information in symbol form and change into appropriate and consistent SI units if required.
v = 1500 m/s
The time taken for the sound to travel there and back is 0.40 s, so
t = 0.20 s
d = ?

Step 2 Use the correct formula.
$$v = \frac{d}{t} \Rightarrow d = vt$$

Step 3 Calculate the answer by putting the numbers into the formula.
d = 1500 × 0.20 = 300 m

ALWAYS REMEMBER TO STATE THE UNIT FOR CALCULATED QUANTITIES.

Unit 3 Properties of waves

Notes

Sound wave characteristics

- The **speed of sound** varies from material to material. A sound wave compresses (pushes together) and rarefies (spreads apart) the particles of the material, and how easily this happens affects the speed of the wave.

- Sound travels fastest in solids because the molecules in a solid are very close together.

- Sound travels second fastest in liquids because the molecules in a liquid are only slightly further apart.

- Sound travels slowest in gases because the molecules in a gas are far apart.

- **Remember**: Sound cannot travel through a vacuum because there are no particles.

- Some examples of the speed of sound in different media are given below.
    - In dry air at 0°C the speed is 330 m/s.
    - In dry air at 20°C the speed is 340 m/s.
    - In water the speed is 1500 m/s.
    - In gold the speed is 3200 m/s.
    - In steel the speed is 5800 m/s.

- Frequency, pitch, wavelength and amplitude are terms associated with sound waves.

- **Frequency** is the number of wavelengths produced per second and its unit is the **hertz** (Hz).

- **Pitch** is how **high** or **low** a sound is and is dependent on frequency.

- **Wavelength** is the distance from any point on one wave to the same point on the adjacent wave. Its unit is the **metre** (m).

❏ **Amplitude** is the maximum disturbance of the wave particles from their equilibrium position. It determines how **loud** or **quiet** a sound is.

Oscilloscope wave traces

❏ The diagrams show graphical representations of sound waves as seen on an oscilloscope screen.

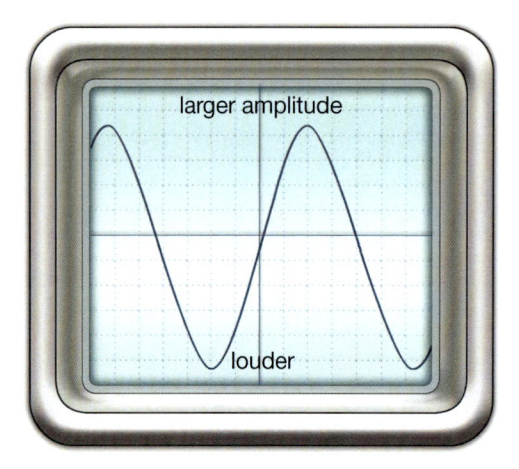

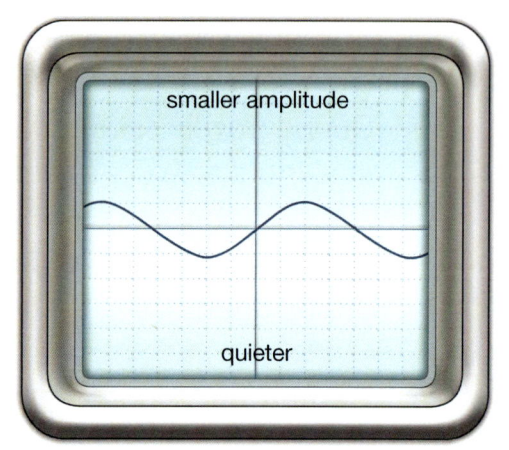

larger amplitude
louder

smaller amplitude
quieter

❏ Notice that the frequency for both the loud sound and the quiet sound is the same in the top traces above. Only the amplitude is different.

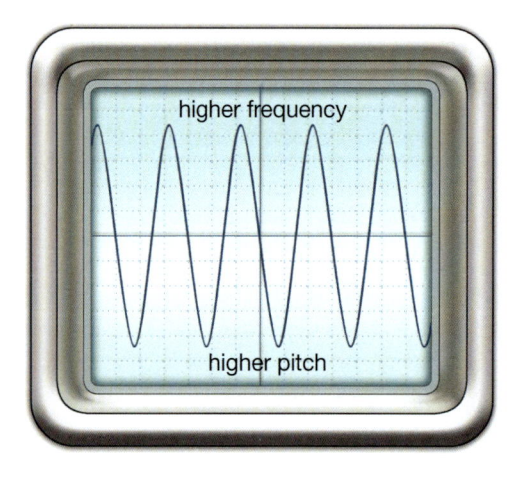

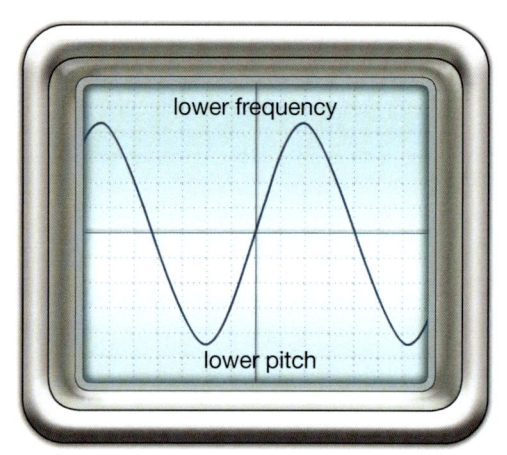

higher frequency
higher pitch

lower frequency
lower pitch

❏ Notice that the amplitude is the same for the high- and low-pitch sounds in the bottom two diagrams. Only the frequency differs.

Remember: Sound waves are longitudinal waves and not transverse waves. The diagrams above are graphical representations.

Unit 4   Electricity and magnetism

Section 4.1 Simple phenomena of magnetism

❏ Properties of magnets:
- They have north (N) and south (S) poles.
- Like poles repel and unlike poles attract.
- Magnets have a **magnetic field** around them – a region in space where their magnetism can affect magnetic materials.
- Magnetic materials contain iron, nickel or cobalt. Steel is a magnetic material; it is an alloy containing iron.
- Magnetic fields can be described using magnetic field lines.

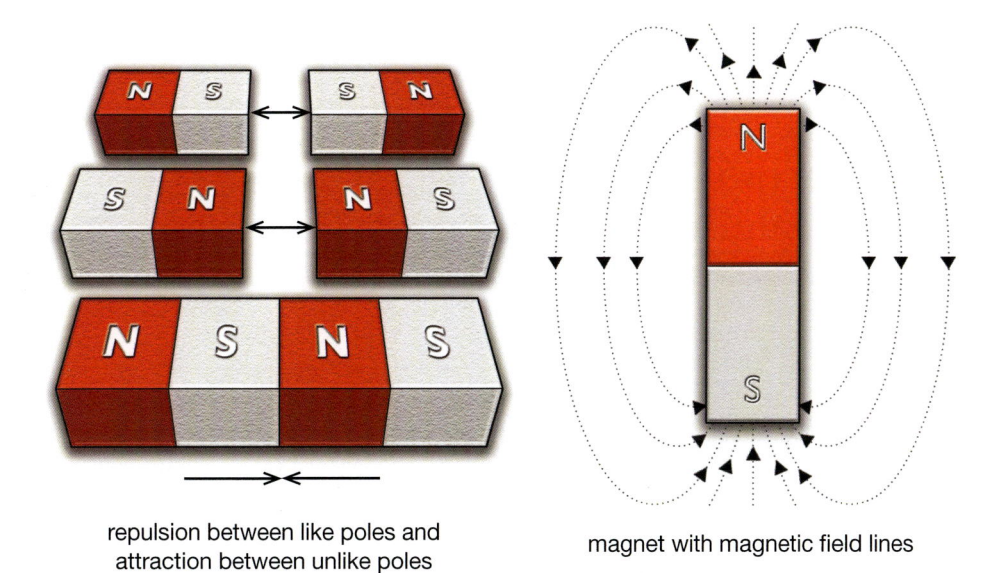

repulsion between like poles and attraction between unlike poles

magnet with magnetic field lines

○ If there is a N pole in the field, then the field at that point is in the direction of the force that acts on the N pole.

❏ The magnetic field lines point from the N pole of the magnet all the way back round to the S pole.

❏ The magnetic field is strongest where the field lines are closest.

○ If two magnets are placed close together their magnetic fields will interact. This interaction produces a force that causes the magnets to move if they are free to do so.

Magnetic fields

❏ The pattern of magnetic field lines can be shown by sprinkling iron filings around a bar magnet. The iron filings arrange themselves in such a way as to show the magnetic field lines, as shown below.

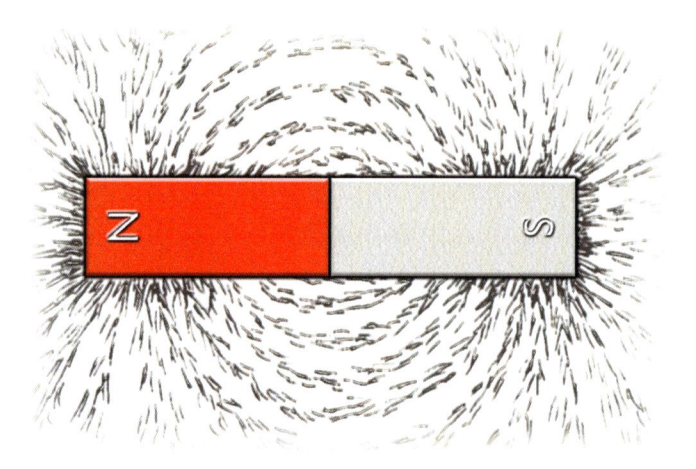

❏ Magnetic field lines can be plotted, and their direction determined, using a plotting compass. The field points from the N pole to the S pole.

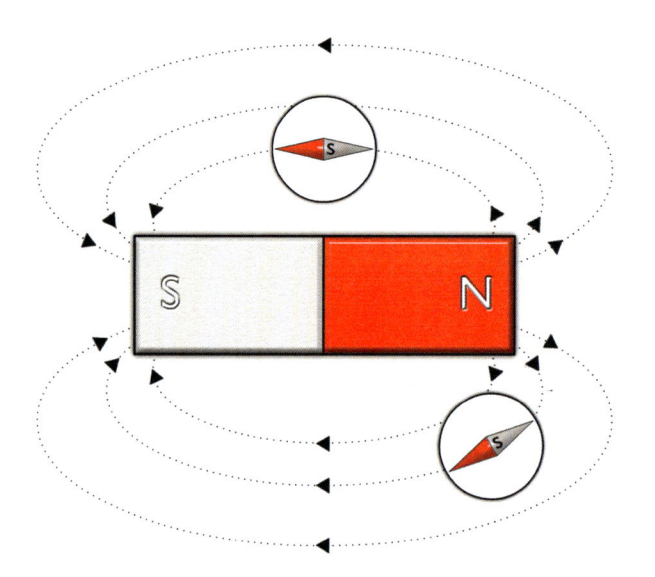

> **Note**
>
> The field lines are always drawn from the N pole to the S pole and they never cross over each other or touch.

Unit 4 Electricity and magnetism

Ferromagnetism

❏ Many magnetic materials are described as **ferrous**. Alloys are made by melting different metallic elements together. The resulting material is ferrous if it contains **iron**. Steel is ferrous. Brass (an alloy of copper and tin) is an example of a non-ferrous metal; it does not contain any iron. Brass is non-magnetic.

Induced magnetism

❏ Magnets can **induce magnetism** in other magnetic materials, i.e. make them into magnets.

The theory of magnetism

❏ Each atom in a magnetic material acts as a small magnet. In an unmagnetised object, the atomic magnets do not line up; they point in random directions. In a magnetised object, the atomic magnets line up in groups called **domains**.

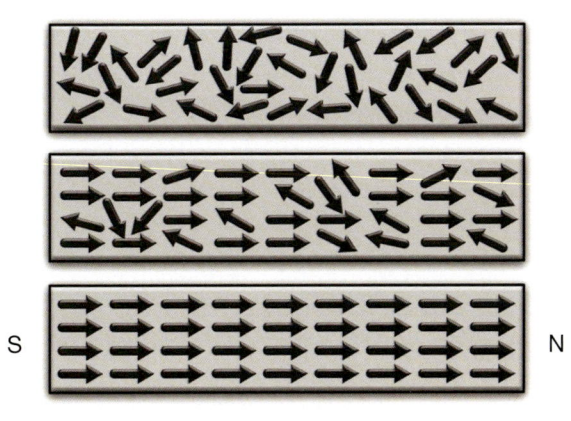

domains not lined up – unmagnetised

some domains lined up – weakly magnetised

all domains lined up – fully magnetised

> **Note**
>
> Magnets always attract unmagnetised magnetic objects. Just because they are magnetic materials does not mean they are magnetised.
>
> The **test** for magnetism is **repulsion**; only two magnets will repel each other.

Methods of magnetisation

❏ **Stroking method**

A bar of steel can be magnetised by stroking it with a bar magnet. The domains become aligned, and a N pole and a S pole are **induced**.

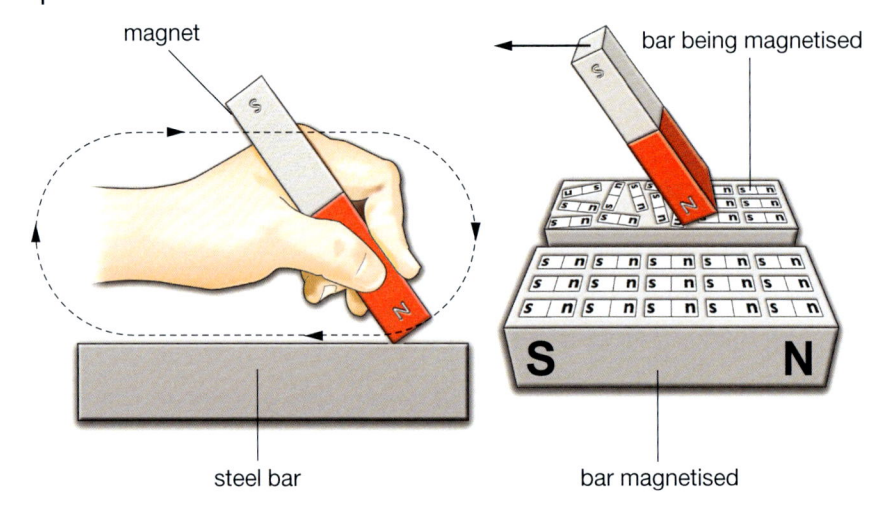

❏ **Hammering**

If a steel bar is placed so that it lies in the direction of the magnetic field lines of a strong magnet and then hammered gently, the domains will begin to line up in the direction of the field. As they do, the steel bar itself becomes magnetised. The effect can be increased by slightly heating the steel bar first.

❏ **Electrical method**

A bar of steel or iron can be magnetised by placing it in a coil of wire (solenoid). Passing a d.c. (direct current) through the wire will magnetise the bar.

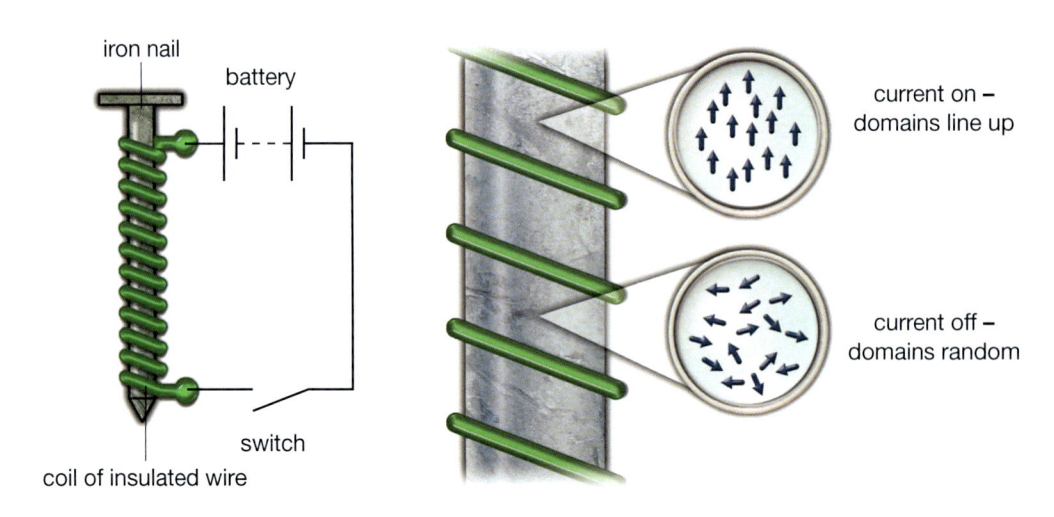

❏ If an iron rod or nail is used as shown in the diagram it is easily magnetised, but when the current is switched off it immediately loses its magnetism. If a steel rod is used instead, it will retain its magnetism when the current is switched off.

Methods of demagnetisation

○ **Hammering**
Hammering vigorously causes the domains to alter their alignment. If there is no external magnetic field to affect the domains they will take up random positions and a magnet will become demagnetised.

○ **Heating**
Heating also causes the domains to alter their alignment. Again, if there is no external magnetic field they will take up random positions.

○ **Electrical method**
A coil of wire is wrapped around the magnet. If a.c. (alternating current) is passed through the wire, this causes the domains to align first one way and then the other. Slowly withdraw the magnet in an east–west direction to demagnetise it. The east–west direction ensures that the Earth's magnetic field does not affect it.

> **Top Tip**
>
> A magnet can be demagnetised by slowly moving it out of a coil of wire (solenoid) whilst alternating current (a.c.) is passing through it.

Hard and soft magnetic materials

- Steel is a **hard magnetic** material. It is difficult to magnetise but once magnetised it is difficult to demagnetise. Steel is often used to make **permanent magnets**.

- Iron is a **soft magnetic** material. It is easy to magnetise, but loses its magnetism easily as well. Iron is often used to make **temporary magnets**.

- The magnetic properties of steel make it a suitable choice to produce permanent magnets such as **bar magnets**.

- Bar magnets have numerous applications, including use as compass needles. They are always magnetised (permanent magnets) and **cannot be switched off**.

- The magnetic properties of iron make it a suitable choice for **electromagnets**. An electromagnet is a coil of wire with a soft iron core. When a current is in the coil the iron becomes magnetised.

- Electromagnets also have numerous applications, such as in door catches, alarm bells and with magnetic inks and paints. They can also be used to lift cars in scrap-yards. They use the magnetic effect of current (page 209) and can **be switched on or off** (temporary magnets).

Section 4.2   Electrical quantities

Electric charge

❏ All matter contains particles called **electrons**, which have an electric charge. When these electrons are taken from the surface of one material and transferred to the surface of another, and stay there, they produce a **static charge** (static means not moving). This branch of Physics is known as electrostatics.

❏ Everyday examples of static charges include:
- rubbing a balloon on your hair and then sticking it to a wall
- rubbing a comb on your jumper and then bringing it near to small pieces of paper; the paper jumps and sticks to the comb
- rubbing polythene and bringing it close to a slow trickle of water; the water is attracted to the polythene.

❏ Some materials are easier than others to charge by rubbing. **Electrical insulators** such as rubber, plastic and glass (non-metals) charge easily.

○ When electrons are added to or removed from an insulator's surface by rubbing, the charges stay on the surface because they are **not free to move** through the material.

❏ Silver, copper and gold are examples of **electrical conductors** (they are metals). Charge passes through them easily.

○ Metals are good conductors because they contain **electrons** that are **free to move**. They are **difficult to charge** by rubbing because the electrons keep moving and the charge flows to earth.

❏ All atoms are made up of three kinds of particles, called **electrons**, **protons** and **neutrons**.

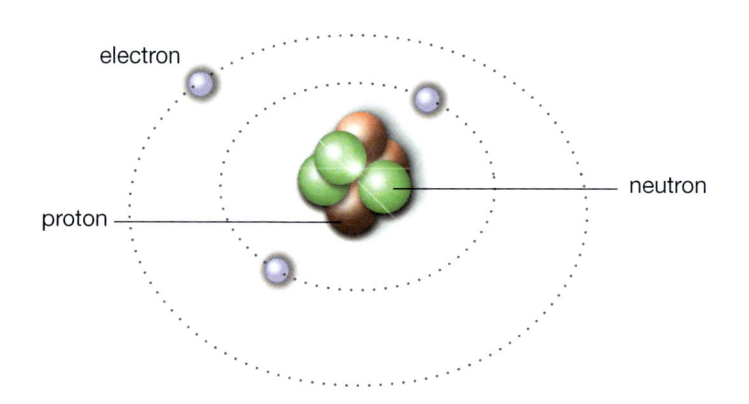

❏ **Protons** and **neutrons** are found in the densest part of the atom known as the **nucleus**, whereas **electrons** are found **orbiting** the nucleus.

❏ **Protons** are **positively** charged, **electrons** are **negatively** charged and **neutrons** have **no charge**.

❏ Atoms usually have **equal** numbers of electrons and protons, making the **resultant charge** of the atom **zero**.

❏ When two different insulators are rubbed together, electrons may be transferred from one insulator to the other. This leaves one insulator positively charged (having lost electrons) and the other one negatively charged (having gained electrons).

❏ **Protons do not** get transferred; it is **only electrons** that get transferred.

> **Note**
>
> Only **electrons** move when charges transfer by rubbing: the **protons do not**.
>
> If an object **gains electrons** it becomes **negatively charged** and if an object **loses electrons** it becomes **positively charged**.

Unit 4 Electricity and magnetism

❏ **Like charges** (+ + or − −) **repel** and **unlike charges** (− + or + −) **attract**.

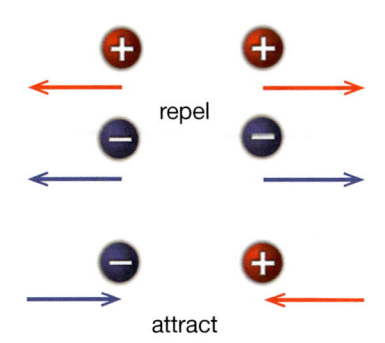

repel

attract

○ Around any charged object there is an **electric field**. A charged particle in this field feels a force towards or away from the charged object, depending on the type of charge on each.

○ An electric field is defined as a region in space in which an electric charge experiences a force.

○ The direction of an electric field at a point is the **direction of the force on a positive charge** at that point. If a positive charge was placed in the fields shown in the diagrams below it would begin to move in the direction of the field lines.

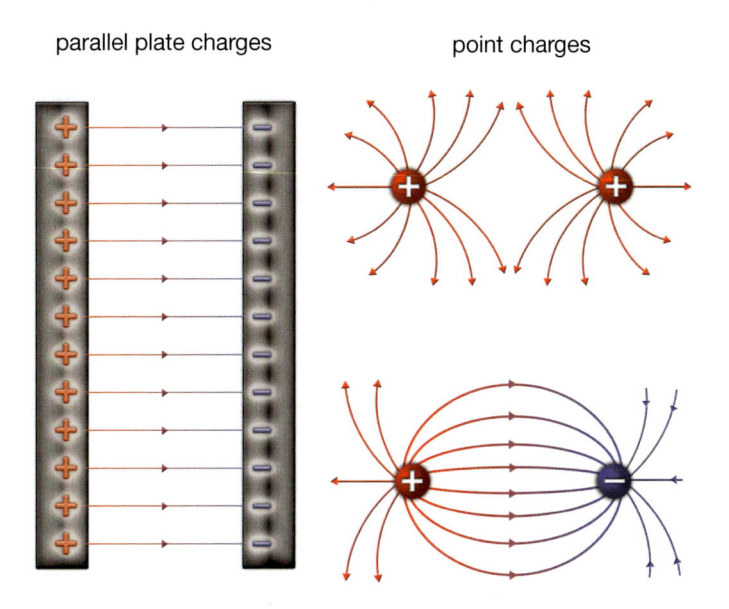

parallel plate charges

point charges

○ The field lines for single point charges and for charged spheres are radial. If the point charge (or sphere) is positive, field lines point away from it; if the charge is negative, field lines point toward it.

> **Note**
>
> Electric field lines are always drawn from positive charges to negative charges. They never cross or touch each other. The electric field is strongest where the electric field lines are closest.

Production of electrostatic charges

❏   Electrostatic charges can be produced in the following ways.

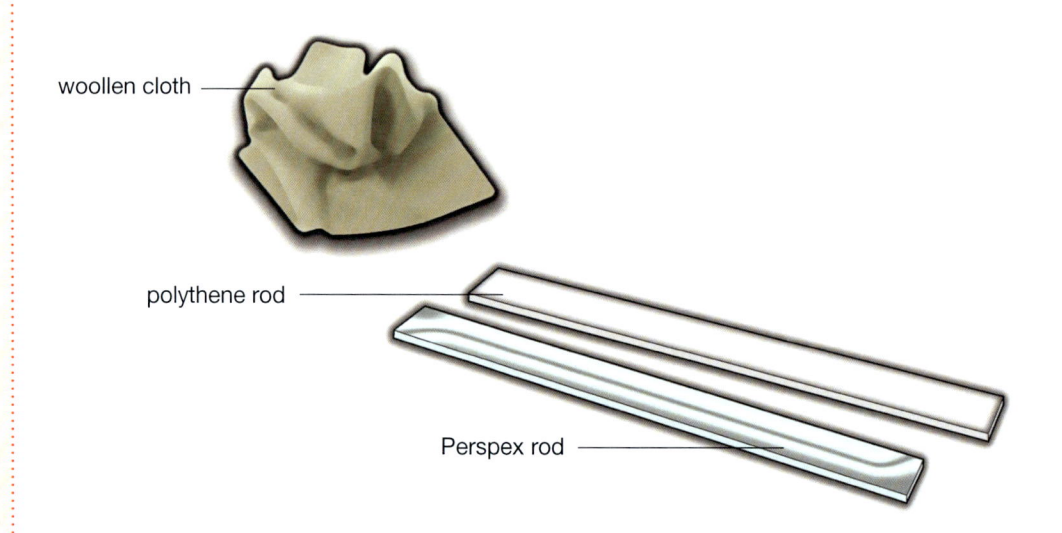

woollen cloth

polythene rod

Perspex rod

1. Rub the polythene rod with the woollen cloth. This transfers electrons from the wool to the polythene, leaving the polythene **negatively charged** (it gains electrons) and the wool **positively charged** (it loses electrons).

2. Rub the Perspex rod with the woollen cloth. This transfers electrons from the Perspex to the wool, leaving the wool **negatively charged** (it gains electrons) and the Perspex **positively charged** (it loses electrons).

> **Top Tip**
>
> The **true test** for charge is **repulsion**, just as with magnets. If you put a positively charged balloon close to an object such as a rod hanging from a thread, and it repels, the rod is definitely charged and has a positive charge.

Unit 4 Electricity and magnetism

Detection of electrostatic charges

❏ A **gold leaf electroscope** can be used to detect if an object is charged. The metal cap is connected to a metal rod and to a strip of very thin gold leaf, which is hinged so it can move.

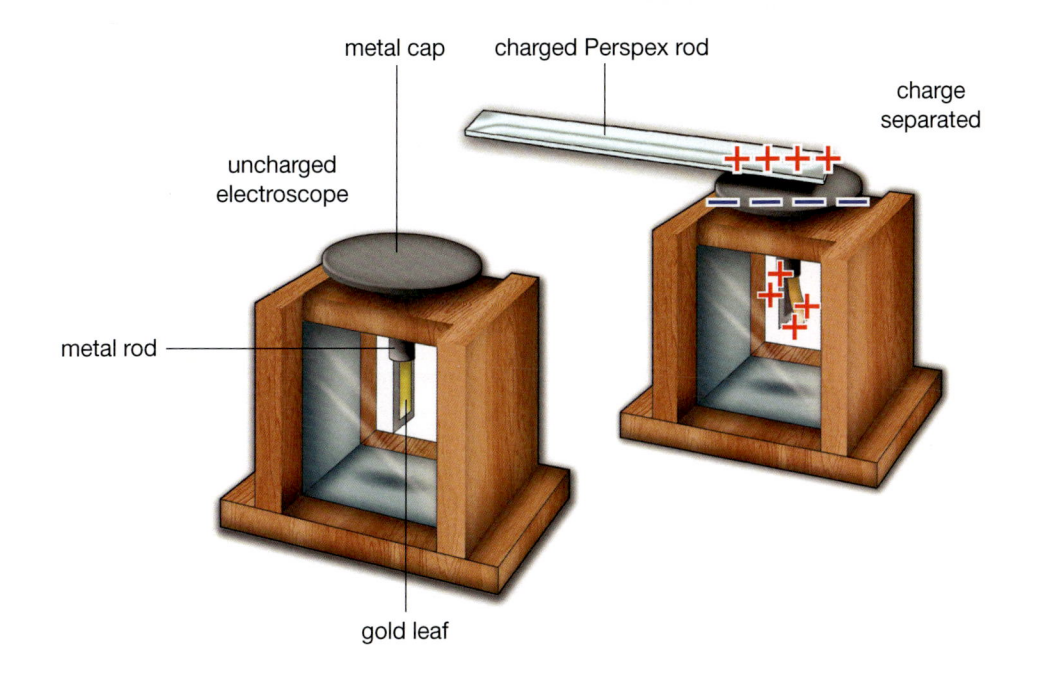

metal cap — *charged Perspex rod* — *charge separated*

uncharged electroscope

metal rod

gold leaf

❏ If you put a positively charged Perspex rod near the cap of the electroscope (**but not touching it**), the gold leaf rises up, away from the metal rod.

○ This happens because the electrons in the electroscope's metal rod and gold leaf are attracted to the Perspex rod. This leaves the cap with a negative charge, and a positive charge on the metal rod and gold leaf (as the free electrons move towards the cap). The positive charges on the metal rod and gold leaf repel each other and so the leaf rises.

Top Tip

Charged objects will sometimes attract **uncharged objects**. If you put a positively charged balloon close to some small pieces of paper it attracts them. The pieces of paper could be either neutral or negatively charged; you can't tell whether the paper is charged or not, so attraction is not a true test.

Charging by induction

○ The following experiment describes how to charge a conducting sphere by **induction**.

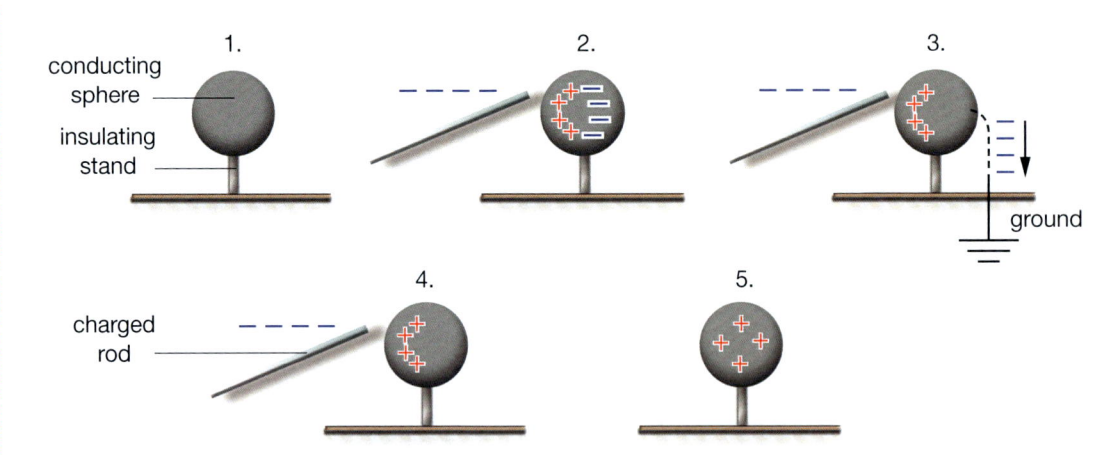

1. Use a neutral metal sphere on an insulating stand.

2. Bring a negatively charged rod towards the sphere. The negatively charged rod causes electrons to be repelled by the negative rod to the right of the sphere, leaving a positive charge on the left.

3. When the sphere is connected to earth (ground), the repelled electrons flow from the sphere to ground through the earth wire, leaving only the positive charges, which are held in place by the negatively charged rod.

4. Disconnect the earth wire, keeping the rod near the sphere.

5. Remove the rod to leave the sphere with an induced positive charge. When the rod is no longer there the positive charge is distributed uniformly over the sphere.

○ The opposite would happen if a positively charged rod was used instead of a negative one.

Unit 4 Electricity and magnetism

Notes

Circuit symbols

It is important to become familiar with the following electric circuit symbols for the rest of this topic. You have to be able to recognise and draw them.

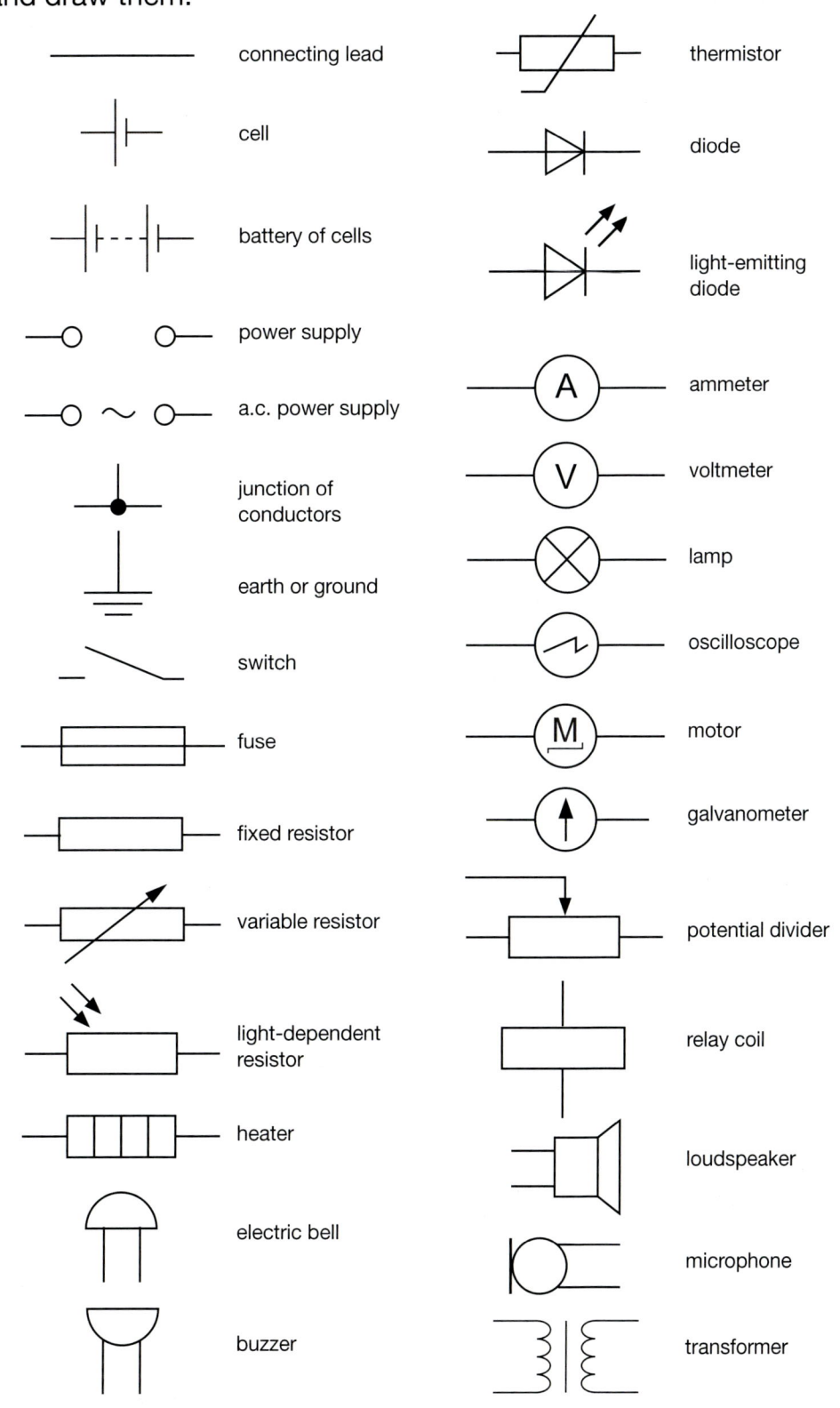

Symbol	Name
—	connecting lead
cell	cell
battery of cells	battery of cells
power supply	power supply
a.c. power supply	a.c. power supply
junction of conductors	junction of conductors
earth or ground	earth or ground
switch	switch
fuse	fuse
fixed resistor	fixed resistor
variable resistor	variable resistor
light-dependent resistor	light-dependent resistor
heater	heater
electric bell	electric bell
buzzer	buzzer
thermistor	thermistor
diode	diode
light-emitting diode	light-emitting diode
ammeter	ammeter
voltmeter	voltmeter
lamp	lamp
oscilloscope	oscilloscope
motor	motor
galvanometer	galvanometer
potential divider	potential divider
relay coil	relay coil
loudspeaker	loudspeaker
microphone	microphone
transformer	transformer

167

Current

- **Electric current** is the **rate of flow** of electric charge.

- **Direct current** (**d.c.**) is the flow of charge in one direction.

- **Alternating current** (**a.c.**) is the flow of charge backwards and forwards. It changes direction many times every second.

- The mains supply (electricity from wall sockets) is always a.c. A battery supply is always d.c.

- Electric current passes in a **conductor** because charges (electrons) are free to move. Current in metals is due to a flow of electrons.

- **Current** has the symbol I, and its unit is the **ampere** (A).

- **Charge** has the symbol Q, and its unit is the **coulomb** (C).

- There is a relationship between **charge**, **current** and **time**. They are related by the formula:

$$Q = It$$

Q = charge (C)
I = current (A)
t = time (s)

Note

The charge on an electron is 1.6×10^{-19} C, which means 1.0 C of charge has 6.25×10^{18} **electrons**, a huge number.

Unit 4 Electricity and magnetism   Notes

○ **Example**

Calculate the current in a wire when 720 C of charge is transferred in 4.0 minutes.

Step 1   List all the information in symbol form and change into appropriate and consistent SI units if required.

$Q = 720\,C$
$t = 4.0\text{ minutes} = 4.0 \times 60\,s = 240\,s$
$I = ?$

Step 2   Use and rearrange the correct formula.

$$Q = It \implies I = \frac{Q}{t}$$

Step 3   Calculate the answer by putting the numbers into the formula.

$$I = \frac{Q}{t} = \frac{720}{240} = 3.0\,A$$

ALWAYS REMEMBER TO STATE THE UNIT FOR CALCULATED QUANTITIES.

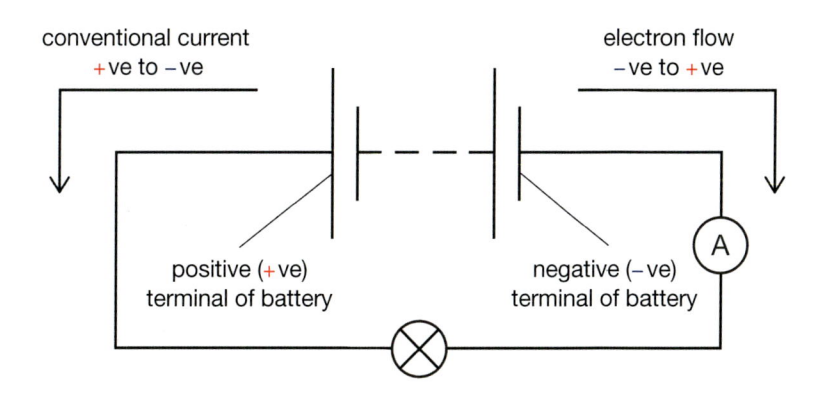

conventional current
+ve to −ve

electron flow
−ve to +ve

positive (+ve) terminal of battery

negative (−ve) terminal of battery

Note

Electrons flow from negative to positive. **Conventional current** is from positive to negative.

Remember: When we talk about current we mean conventional current.

❏ **Current** is measured using an **ammeter**.

❏ An ammeter has **very low resistance** to current and so does not affect the current in the circuit: it simply measures it. An analogue ammeter gives a reading with a needle moving across a scale; a digital ammeter has a numerical reading displayed.

❏ The analogue ammeter below shows a current of 0.50 A.

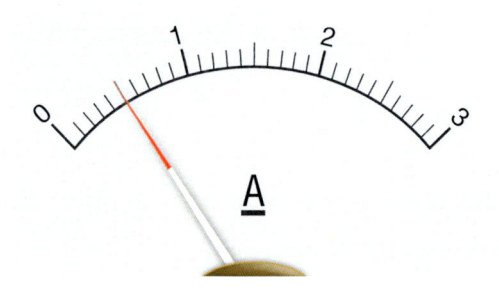

❏ The digital ammeter below shows a current of 1.26 A.

❏ An ammeter is always placed in **series** with other components in a circuit.

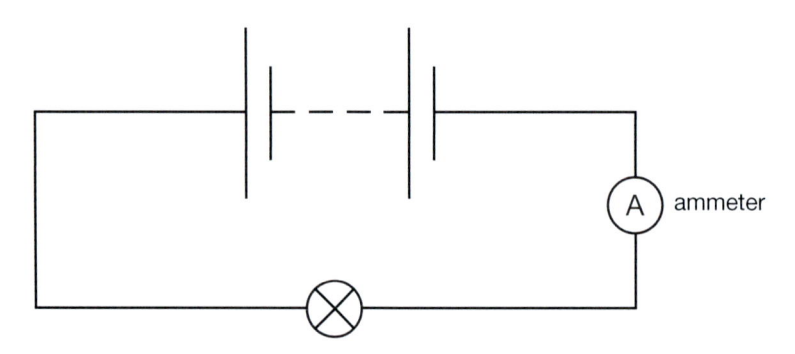

Unit 4 Electricity and magnetism

Electromotive force and potential difference

- **Electromotive force (e.m.f.)** is a measure of how much '**push**' or **energy** a battery or power source can provide to each unit charge in a circuit. Its unit is the **volt** (V).

- The e.m.f. is defined as the amount of **energy** given to **one coulomb** of charge in a circuit (e.g. a **6.0 V** battery gives **6.0 J** to each coulomb of charge).

- **One volt** is defined as one joule per coulomb. **1 V = 1 J/C**

- Charges carry the energy round a circuit to the various components.

- When the charge enters a lamp, for example, the **electrical energy** carried by the charge is transferred as light and heat energy.

- As charge flows through a circuit component it transfers energy, i.e. it has more energy as it starts to flow through the component than when it leaves.

- Voltage is a measure of the difference in electrical energy between two points of a circuit; the bigger the difference in energy, the bigger the voltage. This change in energy is known as the **potential difference (p.d.)** or sometimes the voltage drop or simply 'voltage'.

- The p.d. across a component has the unit **volt** (V).

- A **voltmeter** is used to measure the e.m.f. of a source or the p.d. across a component. It measures the **voltage (potential difference)** between two points.

> **Note**
>
> Voltage is defined as the energy given to each coulomb of charge.
> **1 volt = 1 joule/coulomb or 1 V = 1 J/C**

171

- ❏ A voltmeter can be either analogue or digital. An analogue voltmeter gives a reading with a needle moving across a scale; a digital voltmeter has a numerical reading displayed.

- ❏ The analogue voltmeter below shows a voltage reading of 2.4 V.

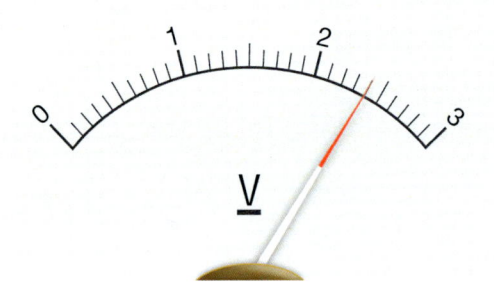

- ❏ The digital voltmeter below shows a voltage reading of 3.32 V.

- ❏ Voltmeters are always connected in **parallel** with the component whose voltage is being measured, as shown in the diagram. They do not affect the circuit. Volmeters have very high resistance.

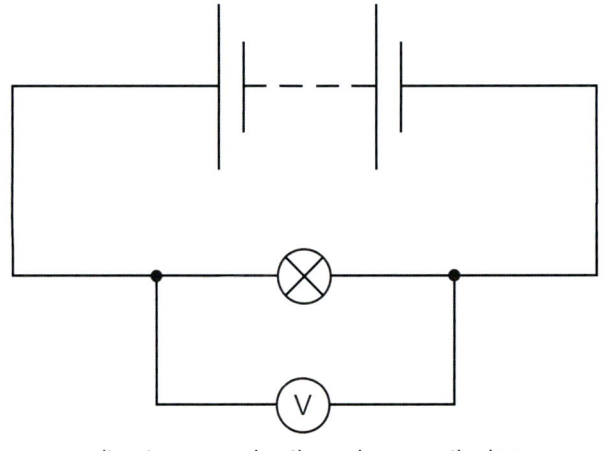

voltmeter measuring the p.d. across the lamp

Unit 4 Electricity and magnetism   Notes

Resistance

❏ All components in an electrical circuit **oppose**, to a small or large degree, the current flowing in them.

❏ **Resistance** can be thought of as **electrical friction**.

❏ Resistance has the symbol **R**, and its unit is the **ohm** (Ω).

❏ When the total resistance in a circuit increases and the voltage of the source remains constant, the current in the circuit decreases.

❏ It can be difficult to visualise what is happening in an electric circuit because electric current is invisible. We can compare an electric circuit with a water circuit.

- Imagine a circuit consisting of water pipes and a pump moving the water round. The pump gives the water kinetic energy just as a battery gives the electrons energy.
- A flowmeter would measure the rate of flow of the water just as an ammeter measures the rate of flow of charge.
- A constriction (narrowing) in the pipe reduces the flow rate (slows the water down) just as a **resistor** reduces the current (slows the charges). Note that the flow rate is reduced both before and after the constriction.

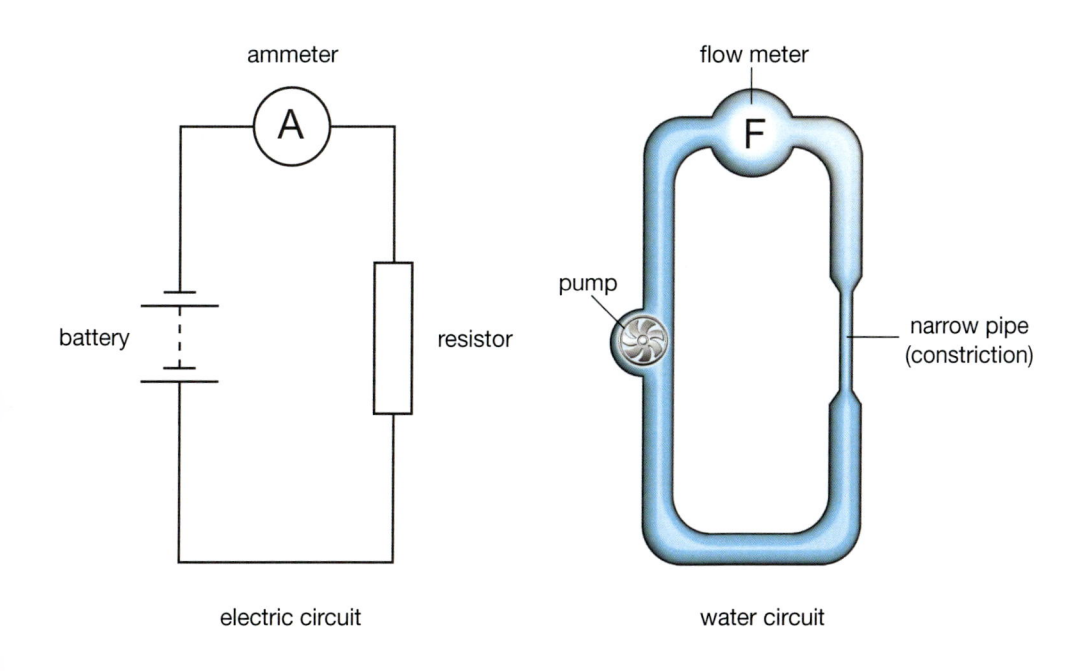

electric circuit   water circuit

173

Factors that affect resistance

❏ The resistance of a wire can be increased by increasing its **length**.

❏ The resistance is proportional to the length.

$$R \propto l$$

Therefore, doubling the length of wire doubles the resistance. Imagine the effect on water flow of doubling the length of the constriction in a water pipe.

❏ The greater the **cross-sectional area A** of a wire, the **more electrons** there are available to carry the charge along a conductor's length and so the lower the resistance.

❏ Doubling the cross-sectional area will halve the resistance **R**. This is known as inverse proportion because the resistance is proportional to the inverse of the area (i.e. 1/**A**).

$$R \propto \frac{1}{A}$$

❏ If you double the **diameter d** of a wire, then its cross-sectional area is four times greater and therefore its resistance **R** is one-quarter of the original resistance.

$$R \propto \frac{1}{d^2}$$

❏ Again, imagine doubling the diameter of a water pipe. Its cross-sectional area will quadruple and the water will be able to pass much more easily. The rate of flow will quadruple, or in other words, the resistance to the flow of water will reduce to one-quarter of its original value.

> **Note**
>
> Resistance is a measurement of a conductor's **opposition** to the **flow of electric current**. It is measured in ohms.

Unit 4 Electricity and magnetism

Example 1

A resistor is made of constantan wire that is 23 cm long and has a diameter of 0.38 mm. Its resistance is measured as 1.0 Ω.

Calculate:
(a) the resistance if the length of the wire is increased to 46 cm
(b) the resistance if the length remains at 23 cm but the diameter is increased to 0.76 mm.

(a) **Step 1** List all the information in symbol form and change into appropriate and consistent SI units if required.

length l_1 = 23 cm
length l_2 = 46 cm
resistance R_1 = 1.0 Ω

Step 2 Recall that resistance is proportional to length.
$R \propto l$

Step 3 The length has doubled, so the resistance will also be doubled.
$R = 2 \times 1.0 = 2.0\,\Omega$

(b) **Step 1** List all the information in symbol form and change into appropriate and consistent SI units if required.

d_1 = 0.38 mm
d_2 = 0.76 mm
R_1 = 1.0 Ω
R_2 = ?

Step 2 The diameter has doubled, so the cross-sectional area will be four times as big and the resistance will be a quarter of the original value.

$$R = \frac{1.0}{4} = 0.25\,\Omega$$

ALWAYS REMEMBER TO STATE THE UNIT FOR CALCULATED QUANTITIES.

○ **Example 2**
Hairdryer X contains a heating element made of resistance wire of length l and cross-sectional area A and has a resistance of R. A second hairdryer Y is made of resistance wire that is three times as long and a quarter of the cross-sectional area of the wire used for X.

Calculate the ratio: $\dfrac{\text{resistance of element in X}}{\text{resistance of element in Y}}$

Step 1 List all the information in symbol form and change into appropriate and consistent SI units if required.

For X: length = l For Y: length = $3l$
         area = A area = $A/4$
      resistance = R resistance = ?

Remember: Because we are finding a ratio we do not need to include units for length and area.

Step 2 Recall how resistance depends on length and cross-sectional area:
$R \propto l$ and $R \propto \dfrac{1}{A}$

Step 3 Considering the length first.
length of wire for Y = 3 × length of wire for X
$R_Y = 3 \times R_X$

Considering the cross-sectional area:
area of Y = one-quarter of area of X
$R_Y = 4 \times R_X$

So, overall:
$R_Y = 3 \times 4 \times R_X$

$\dfrac{\text{resistance of element in X}}{\text{resistance of element in Y}} = \dfrac{1}{12}$

Ohm's Law

- **Potential difference**, **current** and **resistance** are related by the formula:

    $$V = IR$$

    V = potential difference (V)
    I = current (A)
    R = resistance (Ω)

    By rearrangement:

    $$\text{resistance} = \frac{\text{p.d.}}{\text{current}} \qquad R = \frac{V}{I}$$

- This equation **defines the resistance** of a component as V/I and it is constant for metallic conductors provided the temperature is constant.

- So if the **p.d.** across a resistor **doubles**, the **current** in it **doubles**.

- If the resistance of a conductor **remains constant**, a graph of p.d. against current is a straight line. The **gradient** of the line is the resistance of the conductor.

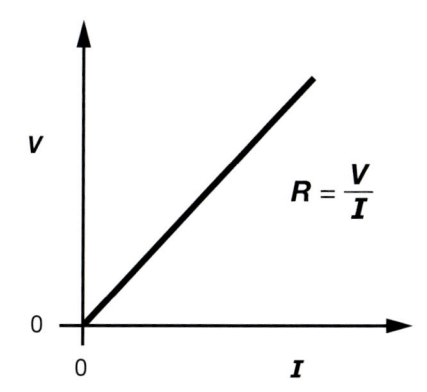

$$R = \frac{V}{I}$$

Note

For a given metallic conductor, V/I is constant provided that its **temperature** is **constant**. This is known as Ohm's Law.

❏ Such a conductor is known as an **ohmic resistor** because it obeys Ohm's Law. The voltage is proportional to the current ($V \propto I$).

❏ If a device has a fixed resistance of R, then you can calculate V or I if you know the other value from $V = IR$.

❏ **Example**
Calculate the current in a resistor that has a 2.0kΩ resistance and is connected across 240V supply voltage.

Step 1 List all the information in symbol form and change into appropriate and consistent SI units if required.

$V = 240\,V$
$R = 2.0\,k\Omega = 2000\,\Omega$
$I = ?$

Step 2 Use and rearrange the correct formula.

$V = IR \quad \Rightarrow \quad I = \dfrac{V}{R}$

Step 3 Calculate the answer by putting the numbers into the formula.

$I = \dfrac{V}{R} = \dfrac{240}{2000} = 0.12\,A$

ALWAYS REMEMBER TO STATE THE UNIT FOR CALCULATED QUANTITIES.

Unit 4 Electricity and magnetism

Determining the resistance of an unknown resistor

❏ The following experiment describes how to determine the resistance of an unknown resistor.

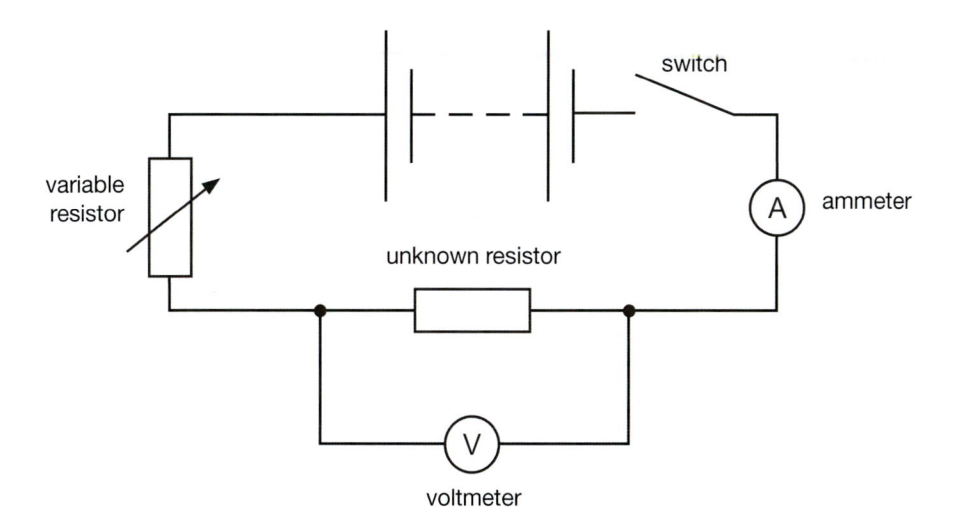

1. Set up the circuit as shown.

2. Vary the current by varying the resistance of the variable resistor.

3. Record the current and voltage.

4. Repeat for several different values of **I** and **V** and construct a table of results.

5. Plot a graph of **V** against **I**.

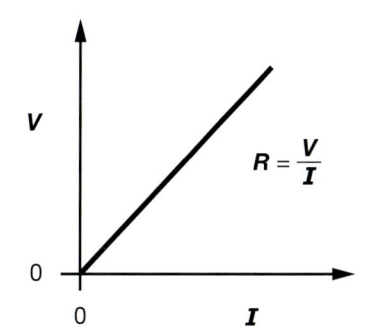

Calculations to be made
Resistance of the unknown resistor.

$V = IR \implies R = \dfrac{V}{I}$

which is equal to the gradient of the **V–I** graph.

Improvements to the experiment
The experiment could be improved by:

1. **switching off** the circuit **between readings** to reduce the heating effect
2. carrying out the experiment at a **lower voltage** to reduce the heating effect.

Assumptions in the experiment
We assume that the changing temperature has **no effect** on the resistor's resistance. A straight-line graph will confirm this.

The effect of heat on resistance

❏ Current in a wire produces a heating effect. This effect is used in toasters, kettles and filament lamps. This happens because **electrons** in the conductor collide with the **atoms** inside the conductor, increasing the resistance. The electrons transfer energy to the atoms, which vibrate faster. This heats the wire.

○ In the graph of current against voltage for the filament lamp shown below, the resistance increases as the voltage increases because the filament becomes hotter. This means that the current does not increase uniformly as it would in an ohmic resistor (see pages 177 and 178).

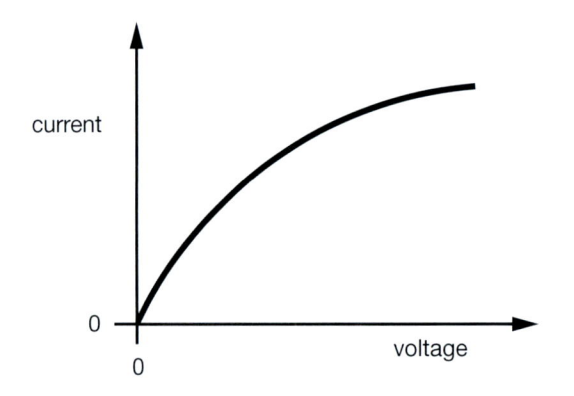

Filament lamp

R increases as voltage increases

Electrical working

❏ Electric circuits transfer energy from the power source or the battery to the circuit components, where it is transferred into useful energy, e.g. a lamp provides light energy.

❏ **Energy** has the symbol *E*, and its unit is the **joule** (J).

❏ Other examples of electrical devices transferring energy include:
   - An electric motor – electrical energy transfers to kinetic energy, which is useful, and heat and sound, which is dissipated to the surroundings.
   - A television set – electrical energy transfers to light and sound, which are useful, and heat, which is dissipated to the surroundings.
   - An ipod – electrical energy transfers to sound energy, which is useful, and heat energy, which is dissipated to the surroundings.
   - An immersion heater – electrical energy transfers to heat energy in the water, which is useful, but some heat energy is dissipated to the surroundings.

❏ All electrical equipment has a **power rating**.

❏ **Power** is the **rate** at which energy is transferred from one form into another (e.g. 2000J of electrical energy being transferred to heat energy per second by a 2000W electric kettle). Power has the symbol *P*, and its unit is the **watt** (W):

   1W = 1J/s

❏ The power rating indicates how much energy is supplied to the device each second, e.g. a **100W** lamp receives energy at a rate of **100J/s**.

❏ Some of the energy is always dissipated to the surroundings, e.g. as well as giving out light, the lamp heats up and the heat energy is 'wasted'.

IGCSE Physics Summarised

○ The **power** of an item of electrical equipment depends on **voltage** and **current**.

$$P = IV$$

P = power (W)
I = current (A)
V = voltage (V)

○ **Example**
The current in a 240 V electric cooker is 8.0 A.
Calculate the power of the cooker.

Step 1 List all the information in symbol form and change into appropriate and consistent SI units if required.

$V = 240\,V$
$I = 8.0\,A$
$P = ?$

Step 2 Use the correct formula.

$P = IV$

Step 3 Calculate the answer by putting the numbers into the formula.

$P = IV = 8.0 \times 240 = 1920\,W$

ALWAYS REMEMBER TO STATE THE UNIT FOR CALCULATED QUANTITIES.

○ The **energy** transferred by an appliance is related to **power** and **time**:

$$E = Pt$$

E = energy (J)
P = power (W)
t = time (s)

○ **Example 1**
Calculate how much energy is transferred by a 100 W lamp in 2.0 minutes.

Step 1 List all the information in symbol form and change into appropriate and consistent SI units if required.

$P = 100\,W$
$t = 2.0\;\text{minutes} = 2 \times 60 = 120\,s$
$E = ?$

Step 2 Use the correct formula.

$E = Pt$

Step 3 Calculate the answer by putting the numbers into the formula.

$E = Pt = 100 \times 120 = 12\,000\,\text{J}$

ALWAYS REMEMBER TO STATE THE UNIT FOR CALCULATED QUANTITIES.

Example 2

A television set is connected to a 240V supply and the current is 0.33A. Calculate the energy transferred if the television is switched on for 1.0 hour.

Step 1 List all the information in symbol form and change into appropriate and consistent SI units if required.

$I = 0.33\,\text{A}$
$V = 240\,\text{V}$
$t = 1.0\,\text{h} = 3600\,\text{s}$
$E = ?$

Step 2 Choose the correct formulae.

$P = IV \qquad E = Pt$

Step 3 Calculate the answer by putting the numbers into the formulae.

$P = IV = 0.33 \times 240 = 79.2\,\text{W}$

$E = Pt = 79.2 \times 3600 = 285\,120\,\text{J}$
$ = 285\,000\,\text{J}$ (to 3 significant figures)
$ = 285\,\text{kJ}$

ALWAYS REMEMBER TO STATE THE UNIT FOR CALCULATED QUANTITIES.

- Since power is related to voltage and current ($P = IV$) and energy is related to power and time, $E = Pt$ can be rewritten as:

  $$\boxed{E = IVt}$$

  E = energy (J)
  I = current (A)
  V = voltage (V)
  t = time (s)

  We simply substitute IV for P to get the formula above.

- **Example**

  Calculate the energy transferred by a 12V hairdryer, running on a current of 0.50A, that is left on for 8.0 minutes.

  Step 1   List all the information in symbol form and change into appropriate and consistent SI units if required.

  $V = 12\,V$
  $I = 0.50\,A$
  $t = 8.0 \text{ minutes} = 8 \times 60 = 480\,s$
  $E = ?$

  Step 2   Use the correct formula.

  $E = IVt$

  Step 3   Calculate the answer by putting the numbers into the formula.

  $E = IVt = 0.50 \times 12 \times 480 = 2880\,J$

  ALWAYS REMEMBER TO STATE THE UNIT FOR CALCULATED QUANTITIES.

- Another expression for electrical power is:

  $$P = I^2R$$

  P = power (W)
  I = current (A)
  R = resistance (Ω)

- The two expressions $P = I^2R$ and $P = IV$ can be shown to be related:

  $P = IV$ but $V = IR$
  Therefore, $P = I(IR)$
  $P = I^2R$

  We simply substitute IR for V to get the formula above.

- **Example**
  Calculate the power dissipated (given out) in a 40Ω pocket torch (flashlight) with a 30mA current in it.

  Step 1   List all the information in symbol form and change into appropriate and consistent SI units if required.

  I = 30mA = 0.030A
  R = 40Ω
  P = ?

  Step 2   Use the correct formula.
  $P = I^2R$

  Step 3   Calculate the answer by putting the numbers into the formula.
  $P = I^2R = 0.030^2 \times 40 = 0.036\,\text{W}$

  ALWAYS REMEMBER TO STATE THE UNIT FOR CALCULATED QUANTITIES.

Section 4.3   Electric circuits

Series and parallel circuits

❏   There is **only one path** for the current in a **series circuit**.

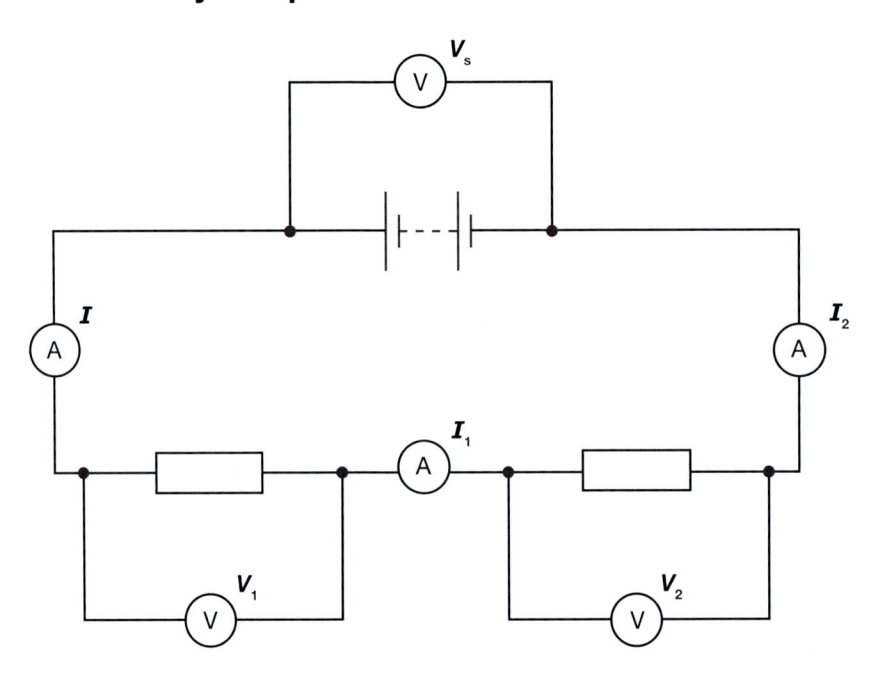

(The various meters are assumed not to be part of the circuit as they are simply there to measure values and have no effect on the circuit.)

❏   The **current** is the **same** at all points in a **series** circuit:

$$I = I_1 = I_2$$ I = supply current as measured by an ammeter.

○   The **supply voltage** V_s, also known as the **electromotive force (e.m.f.)**, is equal to the sum of the voltages (potential differences) across each **individual component** in a series circuit:

$$V_s = V_1 + V_2$$ V_s = supply voltage (e.m.f.)

Note

Current is the same at all points in a series circuit.

The supply voltage (e.m.f.) is shared between components within the circuit.

Unit 4 Electricity and magnetism

○ Sometimes the supply voltage (e.m.f.) is provided by several sources in series. In this case the total e.m.f. is the sum of the individual **e.m.f.s**. For example, if six 1.5V cells are connected in series, the total e.m.f. is 9.0V.

❏ There is **more than one path** for the current in a **parallel circuit**.

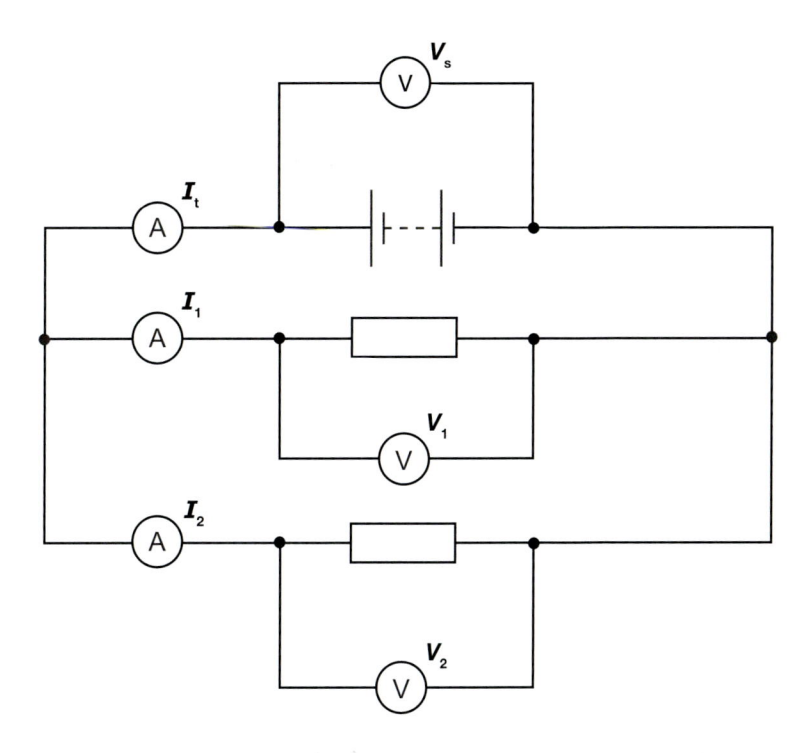

(The various meters are assumed not to be part of the circuit as they are simply there to measure values and have no effect on the circuit.)

❏ The **voltage** (p.d) **across individual** components in a **parallel** circuit is **equal to the supply voltage**:

$$V_s = V_1 = V_2$$

V_s = supply voltage (e.m.f.)
V_1 and V_2 = p.d. across resistors

❏ The **current** in a parallel circuit is larger in the branch containing the battery than in the other branches.

○ The **sum** of the **currents** in the individual parallel branches is **equal to the current drawn from the supply** I_t:

$$I_t = I_1 + I_2$$

I_t = total supply current as measured by the ammeter in series with the battery

❏ The combined total resistance R_t of resistors connected in a **series** circuit can be found by using the following formula:

$$R_t = R_1 + R_2$$

○ The combined total resistance R_t of resistors connected in a **parallel** circuit can be found by using the following formula:

$$\frac{1}{R_t} = \frac{1}{R_1} + \frac{1}{R_2}$$

❏ If two **identical** resistors (each of resistance R) are in parallel then R_t is equal to $R/2$. The total resistance is half of the resistance of one of the two identical resistors. For example, the combined total resistance of two $10\,\Omega$ resistors in parallel is $5.0\,\Omega$.

❏ If two resistors in parallel are not identical then R_t is less than the resistance of either of the two individual resistors (see example below and opposite).

Top Tip

The total resistance of any two resistors in parallel is less than the resistance of either of the individual resistors.

○ **Example**
Calculate the total resistance R_t of the following circuit.

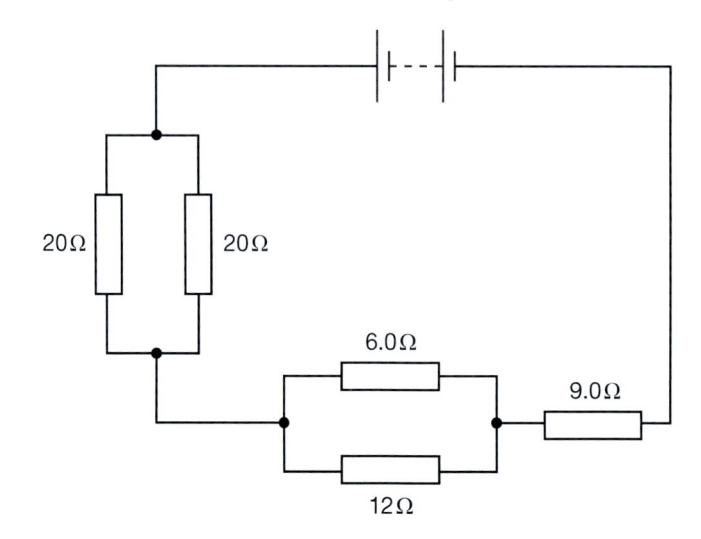

[Circuit diagram: two $20\,\Omega$ resistors in parallel, connected in series with a parallel combination of $6.0\,\Omega$ and $12\,\Omega$ resistors, then in series with a $9.0\,\Omega$ resistor and a battery.]

Unit 4 Electricity and magnetism

Step 1 Calculate the resistance of the pairs in parallel, effectively replacing each pair with a single **equivalent** resistor.

For the two 20Ω resistors in parallel:

$$R_t = \frac{R}{2} = \frac{20}{2} = 10\,\Omega$$

For the 6.0Ω and 12Ω resistors in parallel:

$$\frac{1}{R_t} = \frac{1}{R_1} + \frac{1}{R_2} = \frac{1}{6.0} + \frac{1}{12}$$

Both fractions need to be put over a common denominator:

$$\frac{1}{R_t} = \frac{2}{12} + \frac{1}{12} = \frac{2+1}{12} = \frac{3}{12}$$

Turn both sides upside down (**invert**):

$$\frac{R_t}{1} = \frac{12}{3} = 4.0\,\Omega$$

(In mathematics, this is called finding the **reciprocal**.)

Step 2 Now the circuit becomes:

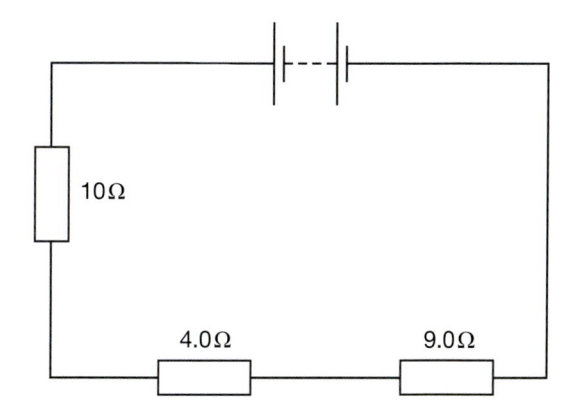

[Circuit diagram showing a battery with a 10Ω resistor, 4.0Ω resistor, and 9.0Ω resistor in series]

The overall total resistance of the circuit can be calculated by using the resistors in series formula:

$$R_t = R_1 + R_2 + R_3 = 10 + 4.0 + 9.0 = 23\,\Omega$$

ALWAYS REMEMBER TO STATE THE UNIT FOR CALCULATED QUANTITIES.

Series circuit with identical lamps

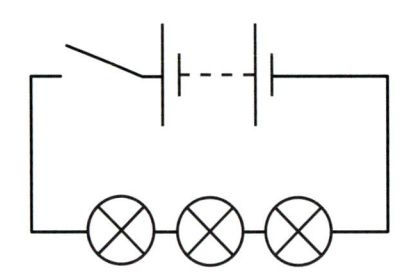

- In a **series** circuit, if one lamp blows the others will **go out** (e.g. some festival tree lights).

- In a series circuit, the lamps **cannot** be switched on and off **independently**. When the switch is open no current can flow in the circuit.

- In a series circuit all lamps **share** the supply voltage (**e.m.f.**) from the battery, so they are **equally lit** when the switch is closed.

- Current is the **same** at **all points** in a series circuit.

- If more lamps are added, the dimmer they all become.

- A practical example of a series circuit is a simple flashlight consisting of a battery, a switch and a lamp.

> **Note**
>
> The **brightness** of the lamps **depends** on **power** ($P = IV$). Lamps in series in a circuit are dimmer than they would be if they were in parallel because there is less voltage per lamp and less current in the lamp because of the higher resistance of the circuit. This in turn causes the power used per lamp to be less.

Parallel circuit with identical lamps

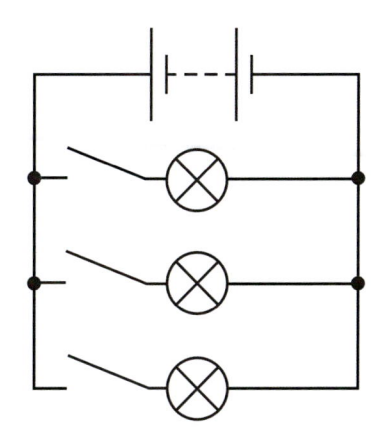

- In a **parallel** circuit, if one lamp blows the others will remain **on**. (Consider the lighting in your house.)

- In a parallel circuit, the lamps **can** be switched on and off independently. If one switch is open it will affect only the lamp in the same branch of the circuit.

- In a parallel circuit, all lamps have the **same voltage** as the **e.m.f.** of the **battery**.

- The total **current**, found by adding the current through each lamp, **adds** up to the **supply** current in a **parallel** circuit.

- If more lamps are added in parallel the brightness will stay the same.

- A practical example of a parallel circuit is the domestic wiring in your home. You can switch the television off without switching off the computer.

> **Note**
>
> **The advantages of parallel circuits are:**
> - When one component such as a lamp goes off the others stay on (because there is more than one path for the current).
> - Each component such as a lamp has the same voltage as the supply voltage.
> - Each component can be switched on and off independently.

IGCSE Physics Summarised

Action and use of circuit components

Potential dividers

❏ **Resistors** in series **share** the supply **voltage (e.m.f.)**.

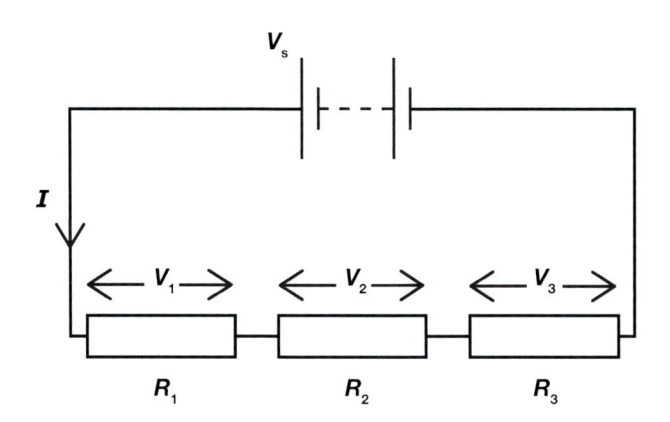

V_s = supply voltage (e.m.f.)

Total resistance: $R_t = R_1 + R_2 + R_3$

Voltage across R_1: $V_1 = IR_1$

but $I = \dfrac{V_s}{R_t}$ ⇨ $I = \dfrac{V_s}{R_1 + R_2 + R_3}$

and therefore $V_1 = \dfrac{V_s}{R_1 + R_2 + R_3} \times R_1$

often written as $V_1 = \dfrac{R_1}{R_t} \times V_s$

Similarly the voltage across V_2 and V_3 can be found as shown on the next page.

Unit 4 Electricity and magnetism

❏ **Example**

Calculate the potential difference (p.d.) across each of the three resistors shown in the circuit diagram below.

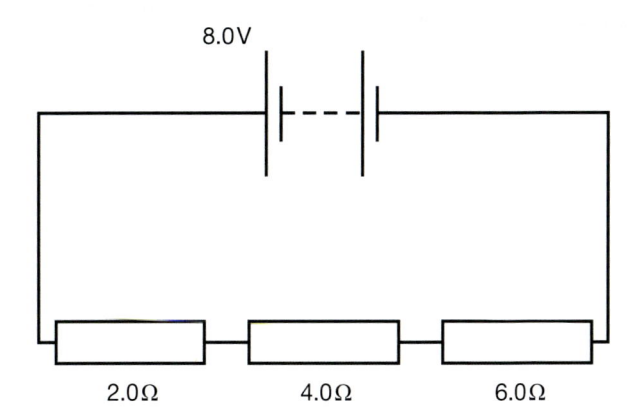

8.0 V

2.0 Ω   4.0 Ω   6.0 Ω

Total resistance = 2.0 + 4.0 + 6.0 = 12 Ω

p.d. across 2.0 Ω: $V = \dfrac{2.0}{12} \times 8.0 = 1.3\,V$

p.d. across 4.0 Ω: $V = \dfrac{4.0}{12} \times 8.0 = 2.7\,V$

p.d. across 6.0 Ω: $V = \dfrac{6.0}{12} \times 8.0 = 4.0\,V$

Note: 1.3 + 2.7 + 4.0 = 8.0 V

This is **same** as the **e.m.f.** because voltage in a series circuit adds up to the supply voltage (e.m.f.) (see page 186).

ALWAYS REMEMBER TO STATE THE UNIT FOR CALCULATED QUANTITIES.

❏ The device shown below can be used as a variable resistor (or rheostat) when it is connected into a circuit using terminals X and Z. It can also be used as a **potentiometer** or **potential divider** when all three terminals are used as shown in the circuit diagram.

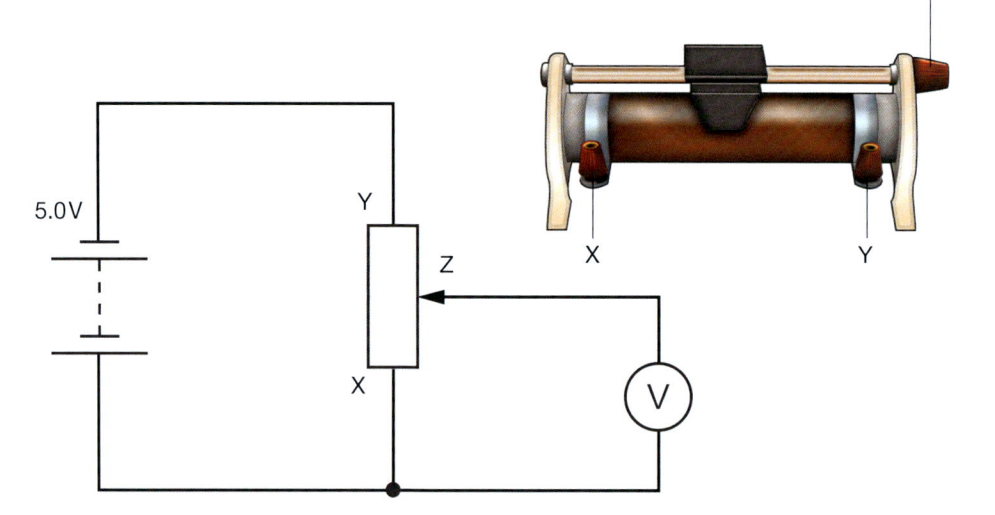

❏ As the slider Z is moved from X to Y, the resistance between X and Z increases, so the voltage across XZ increases. This is the voltage measured by the voltmeter. Therefore the resistance between Z and Y decreases, so the voltage across ZY decreases.

❏ If Z is in the middle of the potentiometer, then the voltage across XZ and the voltage across ZY will be 2.5V each (for the example above).

Other circuit components

❏ Many real-life circuits are used for controlling something.

❏ Circuits with microchips and other **electronic** devices are called **electronic** circuits.

❏ Most electronic circuits operate on **very low current**, although they can control much more **powerful circuits**.

Unit 4 Electricity and magnetism

- All control systems have an input, a **processor** and an output.

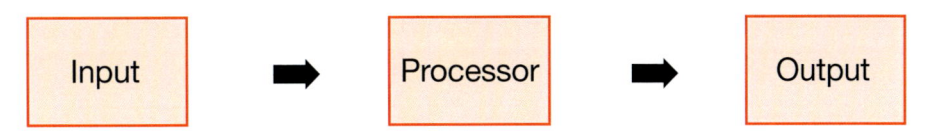

Input ➡ Processor ➡ Output

- An input sensor sends a signal to a processor, which controls the flow of power to an output device.

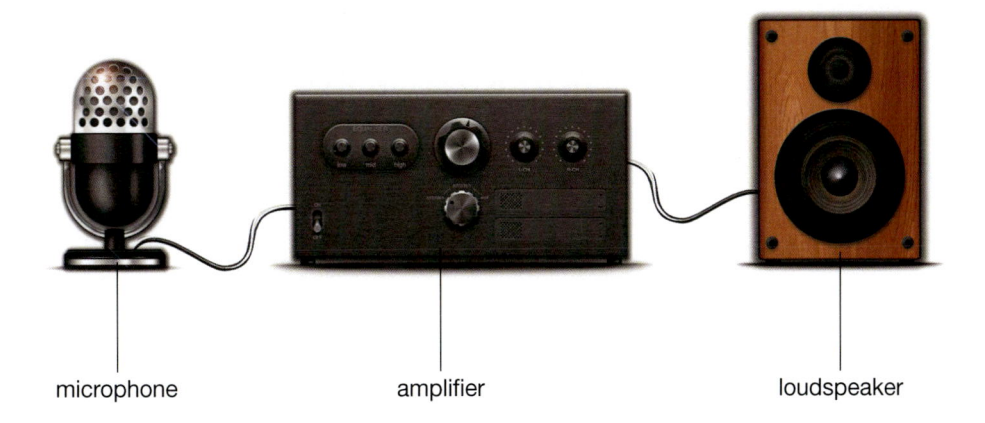

microphone   amplifier   loudspeaker

- **Sound energy** is transferred by the microphone into **electrical energy**, which is then amplified (increased) by the amplifier. The resulting output is a loud noise as **electrical energy** is transferred into **sound again** by the loudspeaker.

- Devices that **transfer** energy from one form to another or electrical signals into some other form of energy are called **transducers**. The components below are all transducers.

Input sensors	Output devices
light-dependent resistor (LDR)	light-emitting diode (LED)
microphone	lamp
thermistor	buzzer
variable resistor	loudspeaker
pressure switch (switch operated by pressing it)	electric motor

❏ A **relay** is an **electromagnetic switch**; it can be used to switch on high-powered circuits. See page 211 for an explanation of how a relay works. The symbol for a relay is:

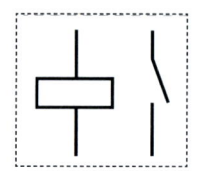

❏ A **variable resistor** controls the current and often the voltage of circuit components. The symbol for a variable resisitor is:

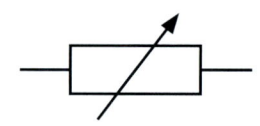

❏ A **thermistor** is a resistor with a resistance that depends on temperature. As the **temperature increases** the **resistance decreases** and vice versa. The symbol for a thermistor is:

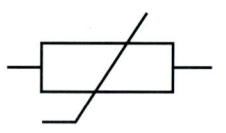

❏ The resistance of a **light-dependent resisto**r (**LDR**), varies according to the amount of light falling on it; as the **light intensity increases** the **resistance decreases** and vice versa. The symbol for a light-dependent resistor is:

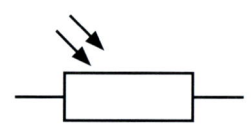

○ A **diode** allows **current** to pass in **one direction** only. The symbol for a diode is:

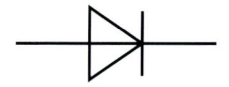

○ A **light-emitting diode** is a diode that glows when current passes through it. The symbol for a light-emitting diode is:

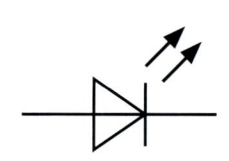

Diodes as rectifiers

○ The process of changing **a.c.** into **d.c.** is known as **rectification**.

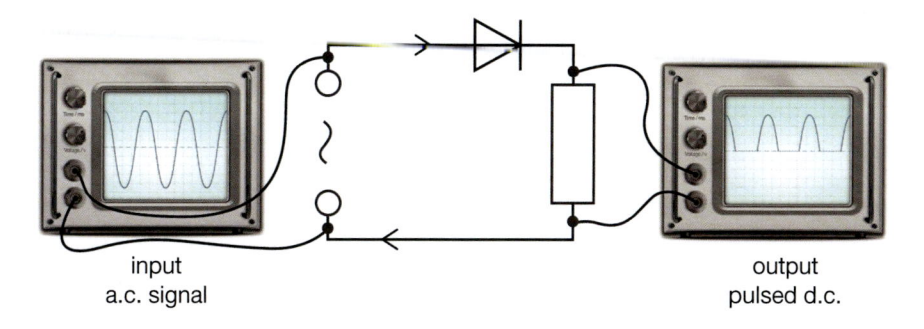

input
a.c. signal

output
pulsed d.c.

○ **Diodes** are used to change a.c. into d.c. in devices that are known as **rectifiers**.

○ Diodes are found in computers, television sets and battery chargers.

○ Diodes have to be **forward-biased** for current to be able to flow. They often need approximately 0.7 V to begin operating.

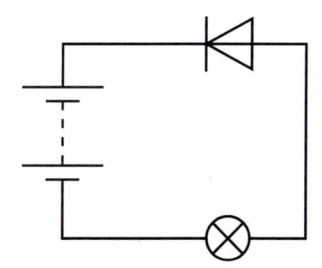

No current
The diode is reverse-biased.
The diode has an infinitely high resistance.

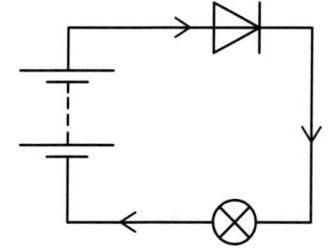

There is a current
The diode is forward-biased.
The diode has a low resistance.

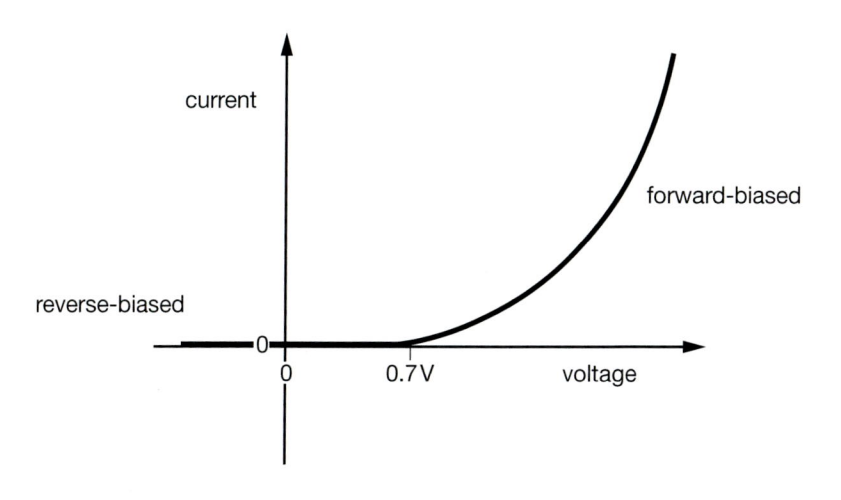

current

reverse-biased

forward-biased

0.7 V   voltage

Light-sensitive switch

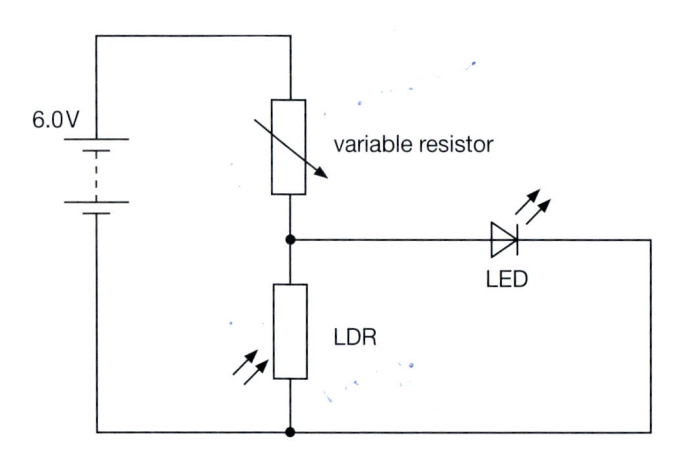

The circuit above switches on the LED when it gets dark. The light from the LED does not fall on the LDR.

How it works
○ The light-dependent resistor (LDR) and variable resistor together behave as a potential divider. The **resistance** of the LDR changes depending on how much light is falling on it.

In daylight:
○ As the intensity of light on the LDR is large, the resistance of the LDR is small.

○ The **potential difference (voltage)** across the LDR is therefore small, so the current through the light-emitting diode (LED) is small and the LED is **off**.

As it becomes dark:
○ As the light intensity incident on the LDR decreases, the resistance of the LDR increases.

○ The potential difference across the LDR increases and consequently it switches the LED **on**.

○ If the **variable resistor** and **LDR** were **swapped**, the lamp would be lit during daylight and be off at night.

Unit 4 Electricity and magnetism

Notes

Temperature-operated alarm

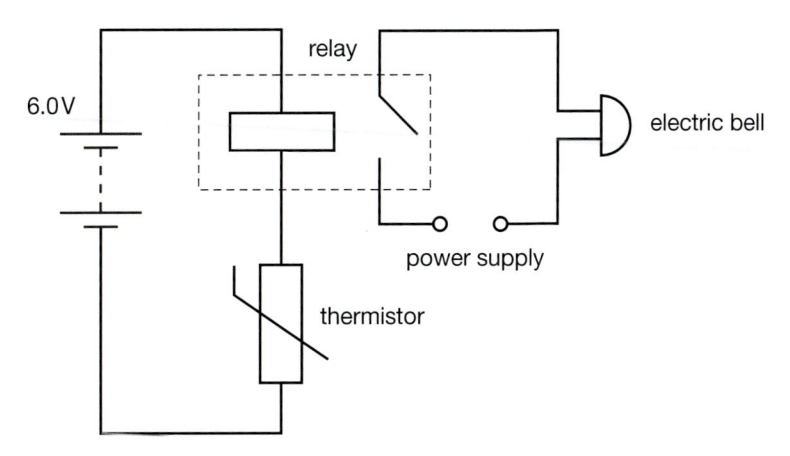

The circuit above causes the alarm to sound when it gets too hot.

How it works

○ The thermistor and relay together behave as a potential divider. The resistance of the thermistor changes depending on the surrounding temperature.

○ A relay is a magnetic switch. If current flows in the coil of the relay it magnetises and causes the switch to close in the second circuit so the bell rings. See page 211 for further explanation of how a relay operates.

When cold:

○ When the temperature is low the resistance of the thermistor is high and therefore the **potential difference** across the thermistor is **high**.

○ The potential difference across the relay coil is **low** and so the current in it is low and its coil does not magnetise enough to close the switch. The bell does not ring.

When warm:

○ When the temperature is high the resistance of the thermistor is low and therefore the **potential difference** across the thermistor is **low**.

○ The potential difference across the relay coil is **high** and so the current in it is large. The coil magnetises and the switch closes and therefore the bell rings.

○ **Example**

The circuit below is designed to light up a warning LED when the temperature falls below a set value.

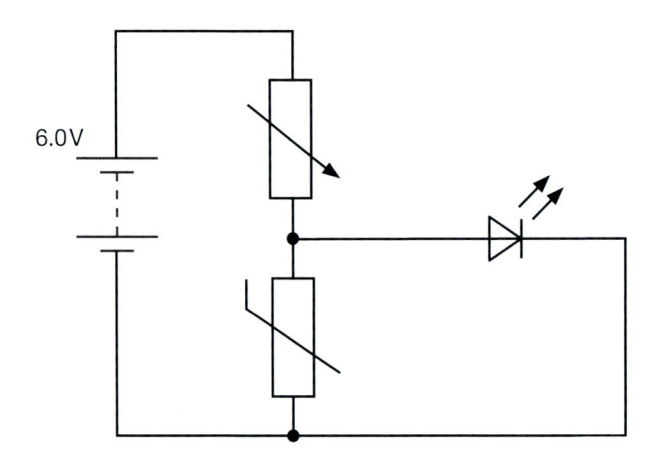

(a) Describe the action of a thermistor.

(b) Explain, with reference to the components in the circuit, why the LED switches on when the temperature falls below a set value.

(c) Describe what happens when:
  (i) the resistance of the variable resistor is changed
  (ii) the variable resistor and the thermistor exchange positions.

(a) The resistance of a thermistor is small when the temperature is high and is large when the temperature is low.

(b) When the temperature decreases, the voltage across the thermistor increases. At a certain temperature, the potential difference across the input of the LED is high enough and so it switches on.

(c) (i) The temperature at which the LED switches on changes.
  (ii) The LED will switch on when it is hot rather than when it is cold.

Section 4.4 Digital electronics

- There are two types of output from electrical or electronic systems:
    - **digital**
    - **analogue**.

- In an analogue output, the signal is **continuously varying**.

- In a digital output, the signal can only be **ON** or **OFF** (**high voltage** or **low voltage**). The output is represented as **1** or **0**.

- Most modern electronic devices (such as cameras, mobile phones, computers, etc.) are digital devices because they process data in the form of numbers (digits).

- To convert an analogue signal to a digital signal an ADC (analogue to digital converter) is used. An ADC includes, amongst other components, a diode to rectify the current.

- On an oscilloscope these signals might look like:

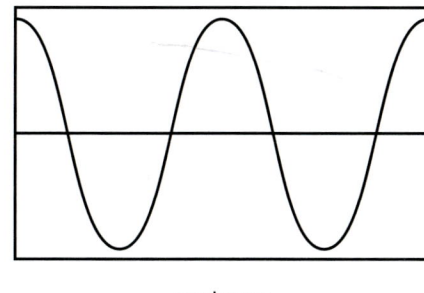

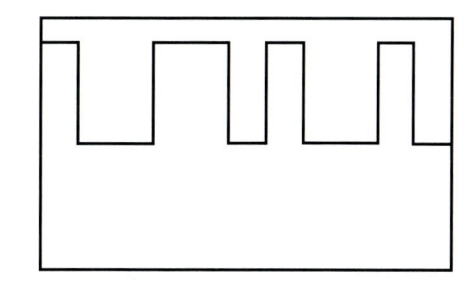

analogue   digital

- It is better to transmit signals in a digital form than in analogue form because **digital signals** can be **amplified without loss in signal quality** and they can carry much **more information**. Telecommunication companies are using digital transmission more and more for these reasons.

Logic gates

- DVD recorders, security lamps, alarm systems and washing machines are just some of the appliances controlled by **logic gates**.

- Logic gates are made up of many electronic components. They are essentially **switches**.

- A logic gate has one or two input terminals, to which a high or low input voltage is connected, and an output terminal whose voltage is determined by the input to the gate.

- The three basic logic gates are the **AND** gate, the **OR** gate and the **NOT** gate:

input A ──┐
          ├─ output
input B ──┘
   AND

input A ──┐
          ├─ output
input B ──┘
   OR

input ──▷o── output
   NOT

- Logic gates make use of two logic numbers:

  high voltage = logic '1' (ON)
  low voltage = logic '0' (OFF)

- **Truth tables** show the outputs for all possible combinations of input.

Unit 4 Electricity and magnetism

○ **AND gate**

An AND gate can be represented by a simple **series** circuit. The circuit below shows you how an AND gate works. Both switches must be closed for there to be a current in the circuit. The circuit itself is **not** an AND gate.

input A	input B	output
0	0	0
1	0	0
0	1	0
1	1	1

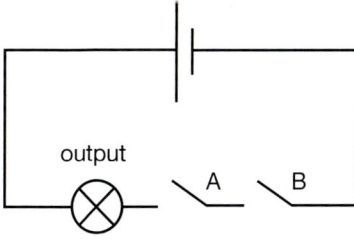

○ **OR gate**

An OR gate can be represented by a simple **parallel** circuit. The circuit below shows you how an OR gate works. There is a current in the circuit if either or both of the switches are closed. The circuit itself is **not** an OR gate.

input A	input B	output
0	0	0
1	0	1
0	1	1
1	1	1

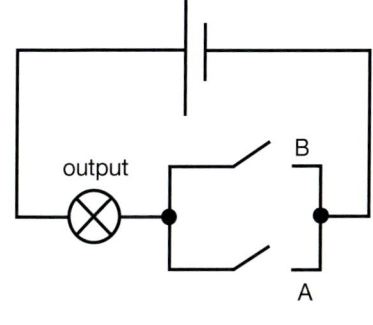

○ **NOT gate**

A NOT gate can be represented by a **simple** circuit with a **short-circuiting switch**. The below circuit shows you how a NOT gate works. There is a current in the lamp when the switch is open. The circuit itself is **not** a NOT gate.

input	output
0	1
1	0

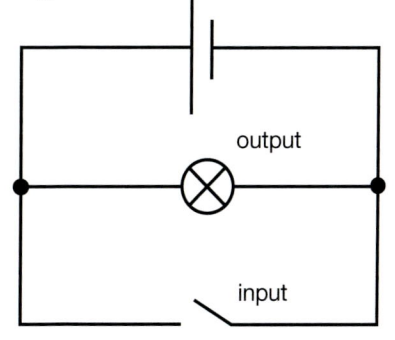

○ Other common gates in use include NAND and NOR gates. A **NAND** gate is a combination of a **NOT** gate and an **AND** gate. A **NOR** gate is a combination of a **NOT** gate and an **OR** gate.

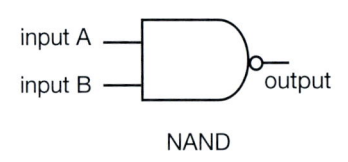

○ *NAND gate*

input A	input B	output
0	0	1
1	0	1
0	1	1
1	1	0

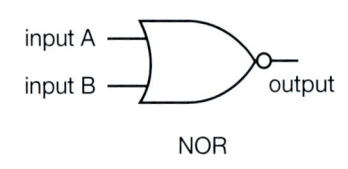

○ *NOR gate*

input A	input B	output
0	0	1
1	0	0
0	1	0
1	1	0

○ Notice from the truth tables how the NAND and NOR gates are the complete **opposite** (inverse) of the AND and OR gates, respectively.

○ More than one logic gate can be combined with others to give the desired effect in a particular situation.

Consider:

- for a washing machine to work the machine must be on **and** the door closed
- a security light will switch on when it is **not** daylight
- a fridge alarm will sound when the door is left open **or** the temperature inside the fridge rises too high.

Unit 4 Electricity and magnetism

○ A truth table can be drawn up for a combination of gates.

Example
Construct a truth table for the following combination of gates.

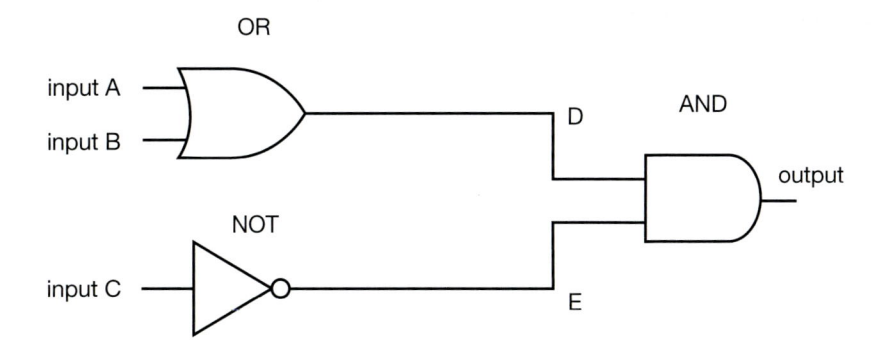

Step 1 Write down all possible combinations of inputs A, B and C into the truth table.

Step 2 Deduce the logic states of D and E for each combination of inputs.

Step 3 Deduce the logic state of the output for each combination of inputs D and E.

input A	input B	input C	input D	input E	output
0	0	0	0	1	0
0	0	1	0	0	0
0	1	0	1	1	1
0	1	1	1	0	0
1	0	0	1	1	1
1	0	1	1	0	0
1	1	0	1	1	1
1	1	1	1	0	0

Top Tip
Remember: You must be able to recognise the logic gate symbols to be able to predict the ouputs and complete the truth table.

Section 4.5 Dangers of electricity

Potential hazards

- **Frayed cables** can lead to the **insulation** around the wires becoming **damaged**, which might lead to the live wire becoming exposed and dangerous. Accidentally touching this would cause an electric shock.

- In addition, exposed cables might touch each other, leading to a short circuit and a large current. **Overheated** cables could lead to a **fire**; this is particularly hazardous due to the fact that much of the wiring in any building is hidden in the walls and underneath floors.

- Having electrical appliances in **damp** conditions (**impure water conducts**) could lead to a person becoming directly connected to the live mains supply, which could lead to serious shock and possible death.

Safety measures

- A **fuse** is a deliberate weak link in a circuit for safety.

- The **fuse** has the symbol: ─┤▭├─

- If there is **too much current**, the fuse wire **melts** and **disconnects the circuit**; it protects the flex (the flexible cable between the plug and appliance) and helps prevent electrical fires.

- The **fuse rating** is always **slightly higher** than the current the appliance usually draws. For example, a 10A appliance should use a 13A fuse. If it used a fuse smaller than 10A, the appliance would not even switch on. The fuse would melt immediately on switching on the appliance.

- A **circuit breaker** is an automatic safety switch. It springs open (**trips**) if too much current flows. This switch can easily be reset once the fault is corrected.

Unit 4 Electricity and magnetism

Example

Determine the appropriate fuse rating for the following appliances. You can choose from a 3A, 5A or 13A fuse.

(a) A 2.5 kW electric kettle operating on a 240V supply.
(b) An 800W electric drill operating on a 240V supply.
(c) A 40W computer operating on a 16V supply.

Step 1   List all the information in symbol form and change into appropriate and consistent SI units if required.

(a) $P = 2.5\,kW = 2500\,W$
     $V = 240\,V$
     $I = ?$

(b) $P = 800\,W$
     $V = 240\,V$
     $I = ?$

(c) $P = 40\,W$
     $V = 16\,V$
     $I = ?$

Step 2   Use and rearrange the correct formula:

$$P = IV \quad \Rightarrow \quad I = \frac{P}{V}$$

Step 3   Calculate the answer by putting the numbers into the formula.

(a) $I = \dfrac{2500}{240} = 10.4\,A$

You would need a 13A fuse.

(b) $I = \dfrac{800}{240} = 3.33\,A$

You would need a 5A fuse.

(c) $I = \dfrac{40}{16} = 2.5\,A$

You would need a 3A fuse.

ALWAYS REMEMBER TO STATE THE UNIT FOR CALCULATED QUANTITIES.

Other safety measures

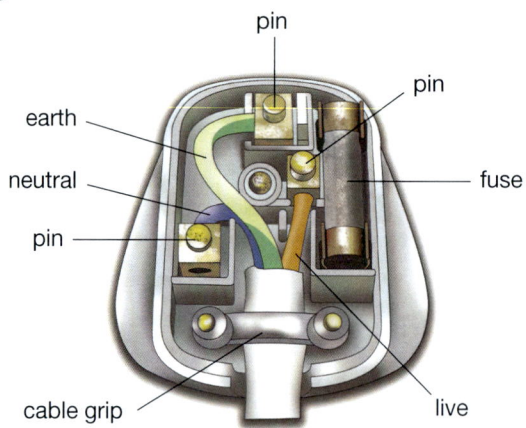

- The **live** wire in a flex is normally brown and is connected to the right-hand pin of a plug. It is linked to the fuse.
  The **neutral** wire is normally blue and is connected to the left-hand pin.
  The green and yellow **earth** wire is connected to the top pin of the plug.
  Colour coding allows us to safely identify wires.
  Colours and plug designs may vary in different countries.

- If the live wire were to become loose and make contact with the **external metal casing** of an appliance, the appliance would become dangerous. The **earth** wire provides the current with an **easy path** from the metal casing of the device to earth. A large current is produced and the **fuse blows**. The circuit is **disconnected**, removing the hazard.

- Some appliances do not need an earth wire as they have a non-conducting (or insulating) casing such as **plastic**. The insulator can never become live. Appliances like this are said to be **double insulated**. Such appliances are marked with this symbol:

Top Tip

Electrical metal wires get hot when there is a current in them. **Thicker** wires can be used to **reduce** this heating effect as the greater cross-sectional area reduces resistance. Using thicker insulation only disguises the heating effect; the wires could still overheat.

Section 4.6 Electromagnetic effects

The magnetic effect of a current

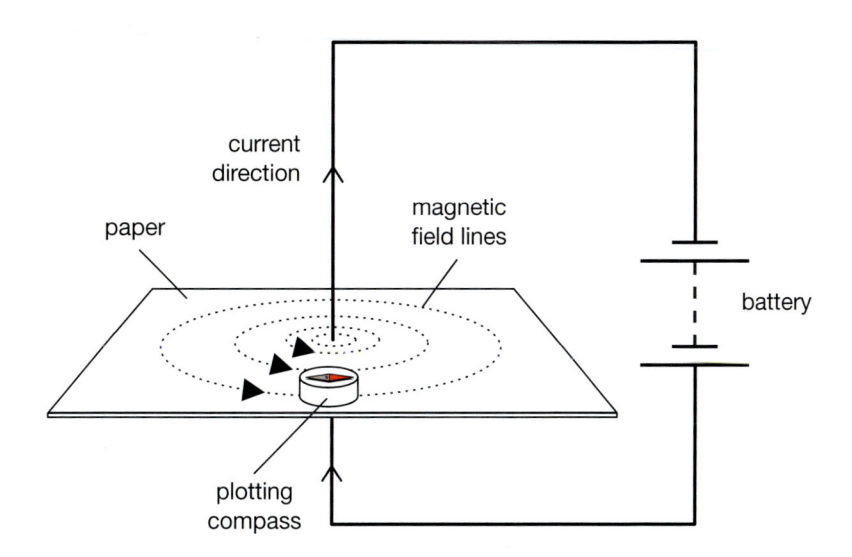

❏ A **magnetic field** is produced around a **current-carrying wire**. The field lines are **concentric circles** around the wire and can be shown using a plotting compass as in the diagram above. The magnetic field is strongest where the field lines are closest.

❏ The current produces a **weak** magnetic field if the current is small.

○ The magnetic field produced by a current-carrying wire has the following features.

- Increasing the current increases the strength of the magnetic field.
- The field is strongest closest to the wire and becomes weaker further away.
- Reversing the direction of the current reverses the direction of the magnetic field.

❏ The magnetic field produced by a single wire carrying a current is not very strong. A stronger field can be produced by a coil of wire (see page 210).

○ The direction of the magnetic field can be found by using the **right-hand grip rule**. Imagine gripping the wire in such a way that the thumb of your right-hand points in the same direction as the current, then your fingers curl around the wire in the direction of the magnetic field.

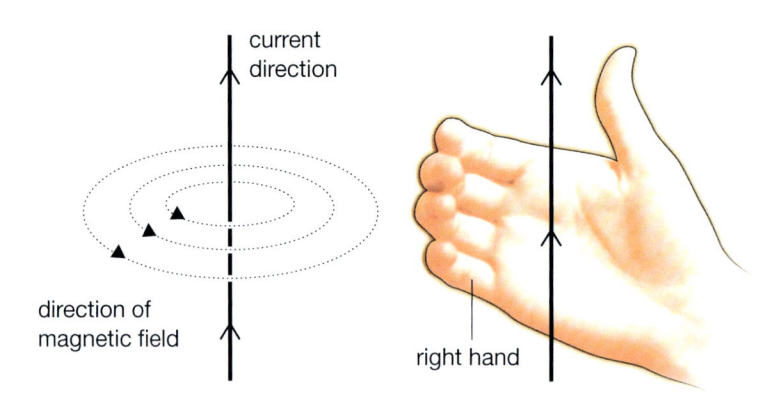

○ In this example, the **magnetic field** is in an **anti-clockwise** direction when viewed from above.

○ In practice, a coil of wire (a **solenoid**) is used to produce a magnetic field, instead of a single straight wire.

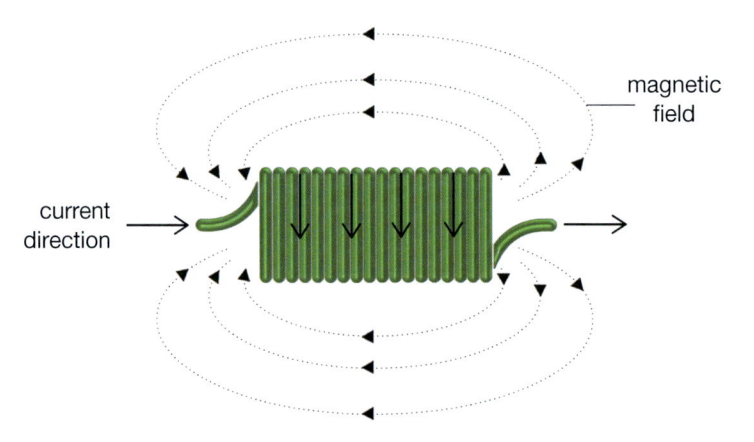

❏ The magnetic field produced by a current-carrying coil or solenoid has the following features.

- The field is similar to the field of a bar magnet and there are magnetic poles at each end of the coil.
- Increasing the current increases the strength of the magnetic field.
- Increasing the number of turns in the coil increases the strength of the magnetic field.
- Having an iron core (shown on the next page) greatly increases the strength of the magnetic field.

Unit 4 Electricity and magnetism

- Reversing the direction of the current reverses the direction of the magnetic field.
- The field is strong close to the coil and weak further away.

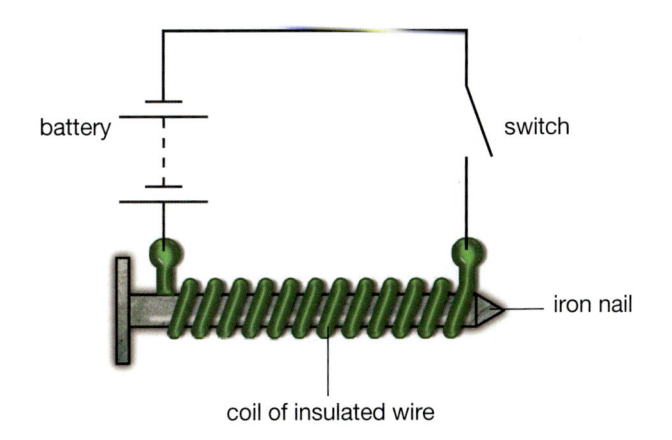

The electric relay

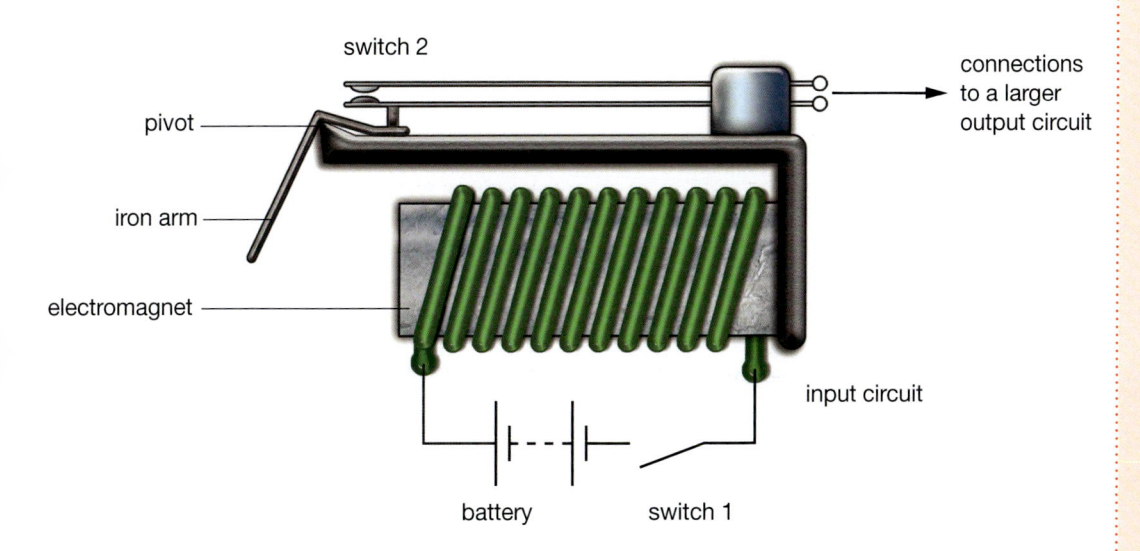

❏ When **switch 1** is **closed** in the diagram above, the input circuit is complete and there is a current in the circuit. The **electromagnet** becomes **magnetised** and it attracts the iron arm. The arm rotates about the pivot and pushes the two contacts of switch 2 together, switching on the output circuit. The output circuit is very often a more powerful circuit, such as the motor circuit for an elevator (lift). The rest of the output circuit is not shown in the diagram above.

❏ The advantage of using a relay is that a **small current** in the input circuit can switch on a **large current** in the **output circuit**.

Force on a current-carrying conductor

❏ When a conductor carrying an electric **current** is placed in a **magnetic field** it will experience a **force**. That force will cause the conductor to move if it is free to do so.

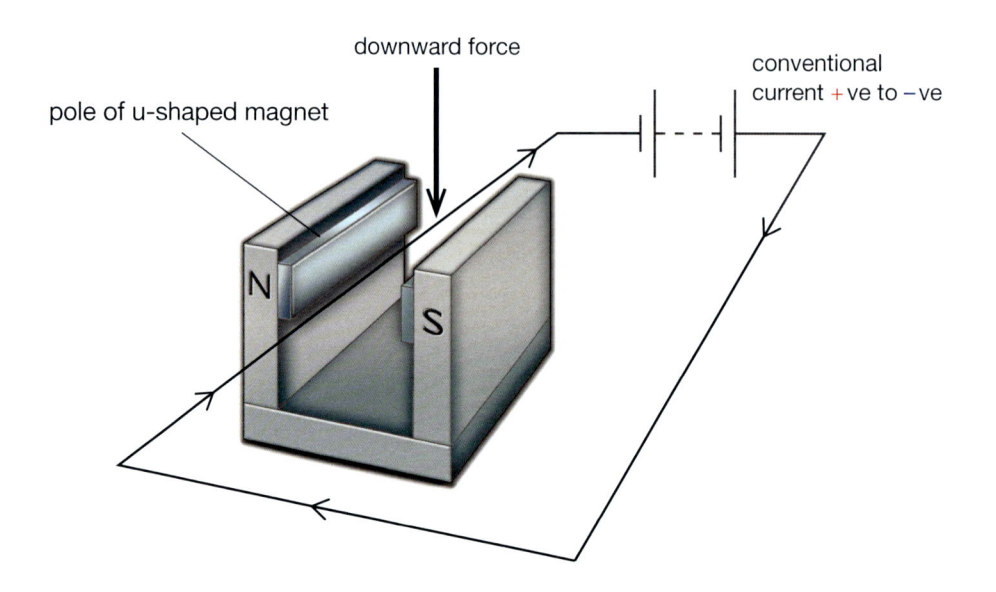

downward force
pole of u-shaped magnet
conventional current +ve to −ve

❏ The diagram shows a wire connected to a battery and placed between the poles of a magnet. When a current flows in the direction shown, the wire is seen to move downwards.

❏ This is because the current-carrying wire has its own magnetic field, which **interacts** with the field of the permanent magnet.

❏ If the current-carrying wire is placed in a magnetic field (whose lines of force are at right angles to the wire) then it will experience a force at right angles to both the current direction and the magnetic field lines.

❏ If either the direction of the current or the magnetic field is **reversed**, the direction of the **force** is **reversed**.

❏ If both the magnetic field and current are reversed, there is effectively no change.

❏ This effect is used in motors.

Unit 4 Electricity and magnetism

❏ The force can be **increased** by:
- increasing the current in the wire
- using a stronger magnet
- increasing the length of wire in the magnetic field.

> **Note**
> - If there is current in a wire in the presence of a magnetic field, a force is produced.
> - This is because the current-carrying wire has its own magnetic field, which interacts with the field of a permanent magnet.

○ **Fleming's left-hand rule – motors**

Fleming's left-hand rule can be used to predict the direction of the force produced in d.c. motors.

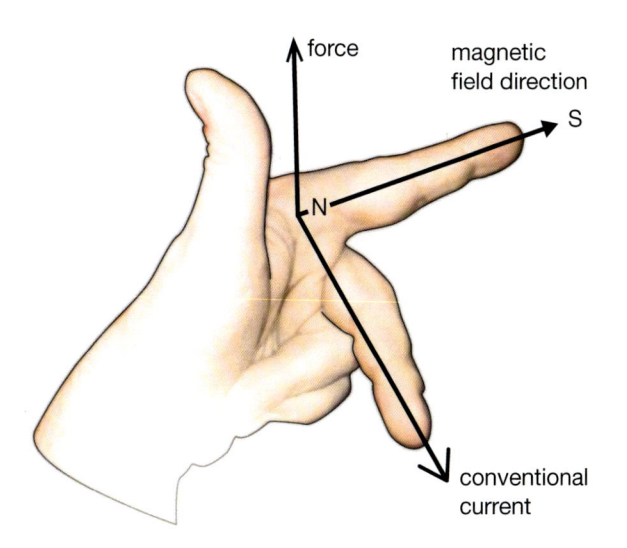

The directions of the current, magnetic field and force are related to each other as shown.
All three fingers have to be at **right angles** to one another.
A tip to help you remember what each finger represents is:

- *f*irst finger-*f*ield
- se*c*ond finger-*c*urrent
- *th*umb-*th*rust

The d.c. motor

❏ A d.c. motor makes use of direct current (current in one direction).

❏ If there is a current in a coil placed between **fixed magnets**, forces act on the coil, producing a **turning effect**.

○ Each end of the coil is connected to one half of a **split-ring commutator** against which **carbon brushes** press. The coil is made of copper wire as it is a good electrical conductor.

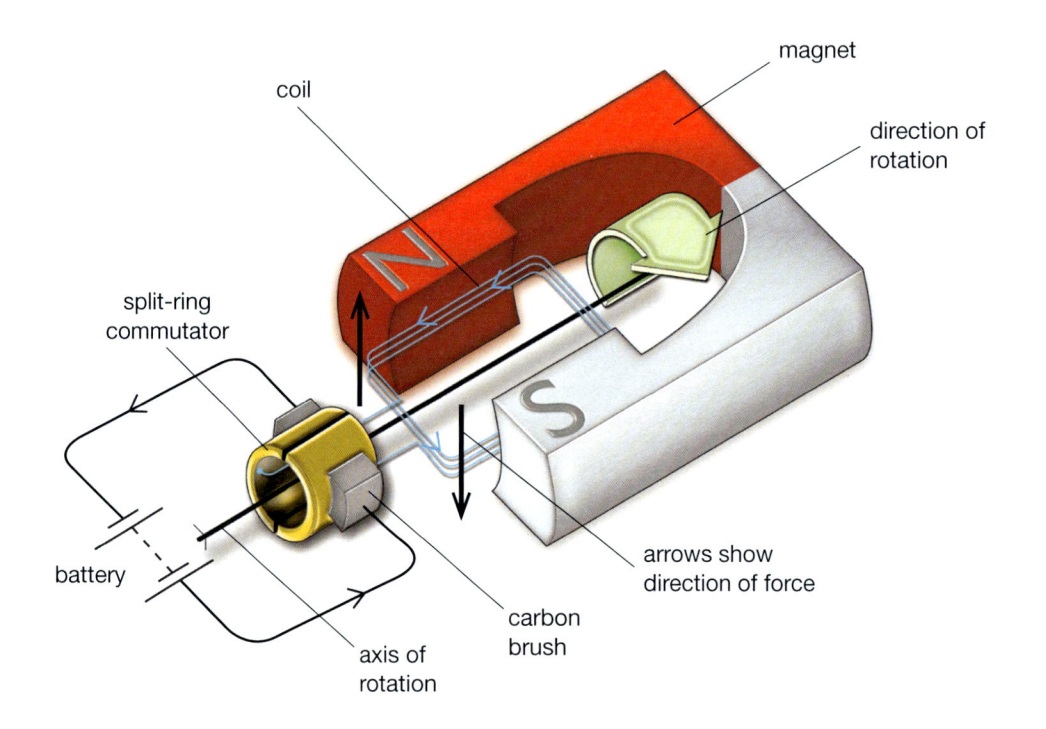

How it works

○ The battery supplies a current to the coil via the split-ring commutator.

○ Because the coil then has a current in it, it also has a magnetic field associated with it.

○ This magnetic field interacts with the magnetic field of the permanent magnet.

- This causes an upward force on one side of the coil and a downward force on the other.

- The direction of each force is given by Fleming's left-hand rule.

- The **split-ring commutator** changes the direction of the current in the coil **every half-turn**, because as the coil rotates the commutator rotates.

- Therefore, the turning force keeps acting in the same direction as the coil rotates, keeping the coil rotating.

- The turning effect, and therefore the speed of revolution of the motor, can be increased by:
    - increasing the number of turns in the coil
    - using a stronger magnet
    - increasing the current.

- In an electric drill, for example, the rotating coil causes a drill bit to rotate, which drills a hole in wood, plastic, metal or stone.

Forces on charged particles in a magnetic field

○ The diagram below shows an electron gun inside a vacuum tube. A beam of electrons is fired towards the fluorescent screen at the front. A spot of light appears where the beam hits the screen.

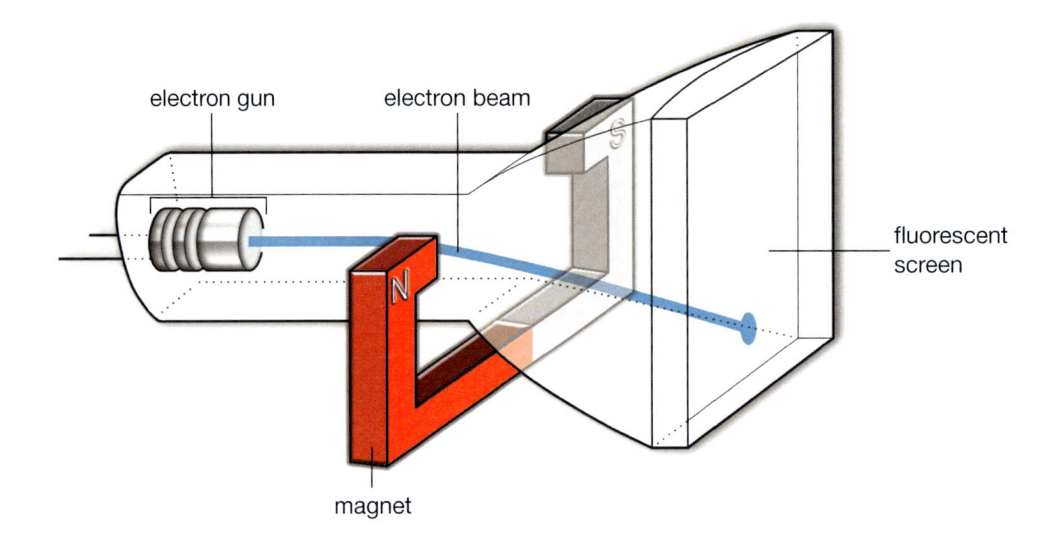

○ A magnet is placed against the outside of the tube and the spot on the screen moves downwards. This is because the direction of the magnetic field is from the N pole to the S pole and at right angles to the direction of movement of the electrons. The electrons are negatively charged, so as they move they constitute an electric current in the opposite direction to their motion (see page 169).

○ Applying Fleming's left-hand rule shows that the electrons experience a force downwards. Hence, the spot on the screen moves downwards.

Electromagnetic induction

❏ Michael Faraday was the first person to generate electricity from a magnetic field using **electromagnetic induction**.

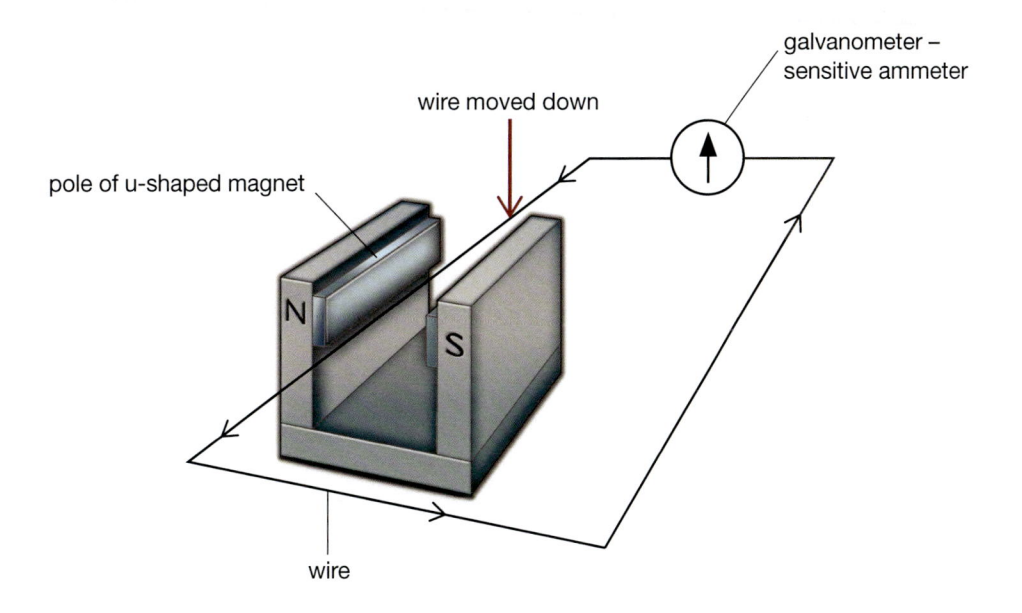

galvanometer – sensitive ammeter

wire moved down

pole of u-shaped magnet

wire

❏ When a wire is moved through a magnetic field a small electromagnetic force (e.m.f.) is produced as the wire cuts the magnetic field lines. This is known as an **induced e.m.f.** If the wire is part of a complete circuit, a current is produced in the wire by the induced e.m.f.

❏ The diagram above shows a wire being moved down between the poles of a magnet and 'cutting' the magnetic field lines. The same effect could be achieved by moving a magnet near a conductor. The galvanometer will detect a current in the conductor if it is part of a complete circuit. The maximum current will be observed when the wire cuts the magnetic field at right angles.

❏ The induced **e.m.f.** can be **increased** by:
- moving the wire faster
- using a stronger magnet to increase the magnetic field
- increasing the length of wire cutting the magnetic field. (This is what happens when a bar magnet is pushed in and out of a coil of wire.)

○ When the induced e.m.f. produces a current its direction **opposes** the effect causing it. In other words, the magnetic field associated with the current caused by the induced e.m.f. opposes the magnetic field produced by the magnets.

○ There is **no e.m.f.** and **no current** if the wire is **moved parallel** to the **magnetic field**. The wire must **cut** the magnetic field lines in order for an e.m.f. or current to be induced.

> **Note**
> - A conducting wire must cut the magnetic field lines in order for an e.m.f to be induced.
> - This e.m.f will be maximum when the field lines are cut at right angles.

○ **Fleming's right-hand rule – generators**
This rule is used to indicate the direction of a current caused by an induced e.m.f.

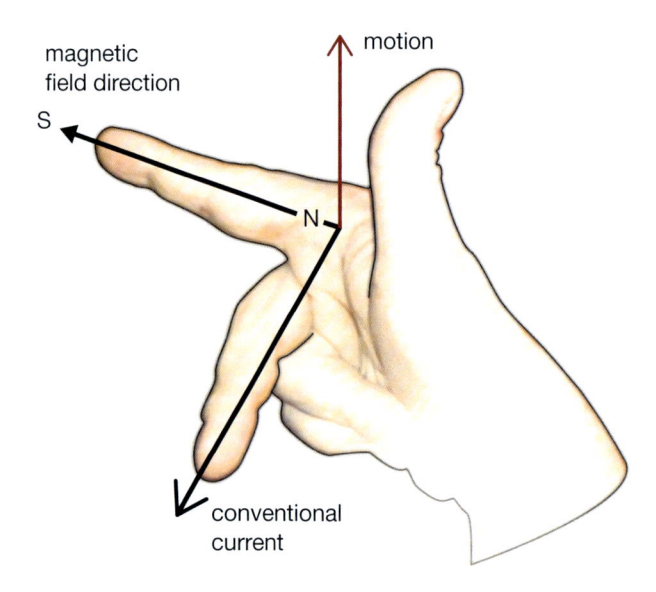

The directions of the current, magnetic field and motion are related to each other as shown.
All three fingers have to be at **right angles** to one another.
A tip to help you to remember what each finger represents is:

- **f**irst finger-**f**ield
- se**c**ond finger-**c**urrent
- thu**m**b-**m**otion

Unit 4 Electricity and magnetism

Induced current in a coil

❏ Electromagnetic induction can also occur in a coil. The wire may be stationary with the magnet moving.

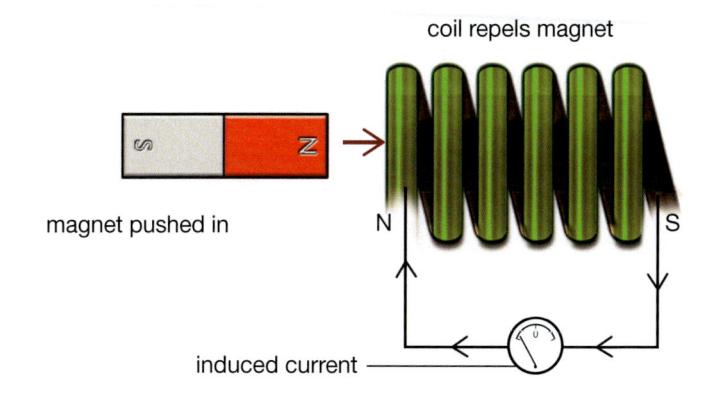

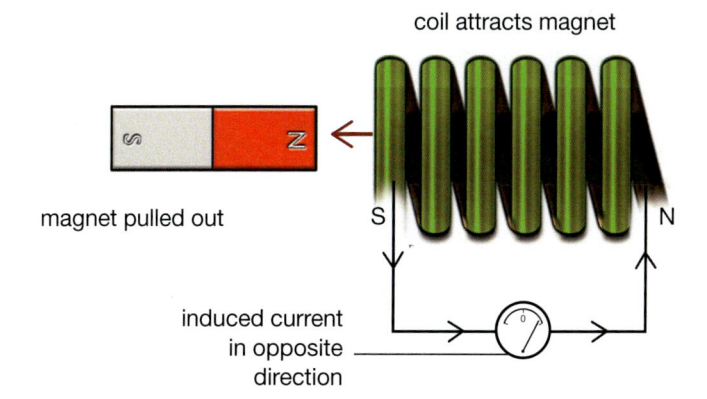

○ The direction of the current opposes the change causing it, whether the magnet is being pushed in or pulled out. This is Lenz's Law.

Remember: no movement = no current

❏ If the magnet is moved in and out several times, the needle on the galvanometer will indicate that the current changes direction (or alternates).

❏ The induced e.m.f. (and the current) can be **increased** by:
- moving the magnet faster
- using a stronger magnet to increase the strength of the magnetic field
- increasing the number of turns in the coil.

The a.c. generator

❑ An **a.c.** generator produces **alternating** (backwards and forwards) e.m.f. and current.

Remember: The term a.c. stands for 'alternating current'.

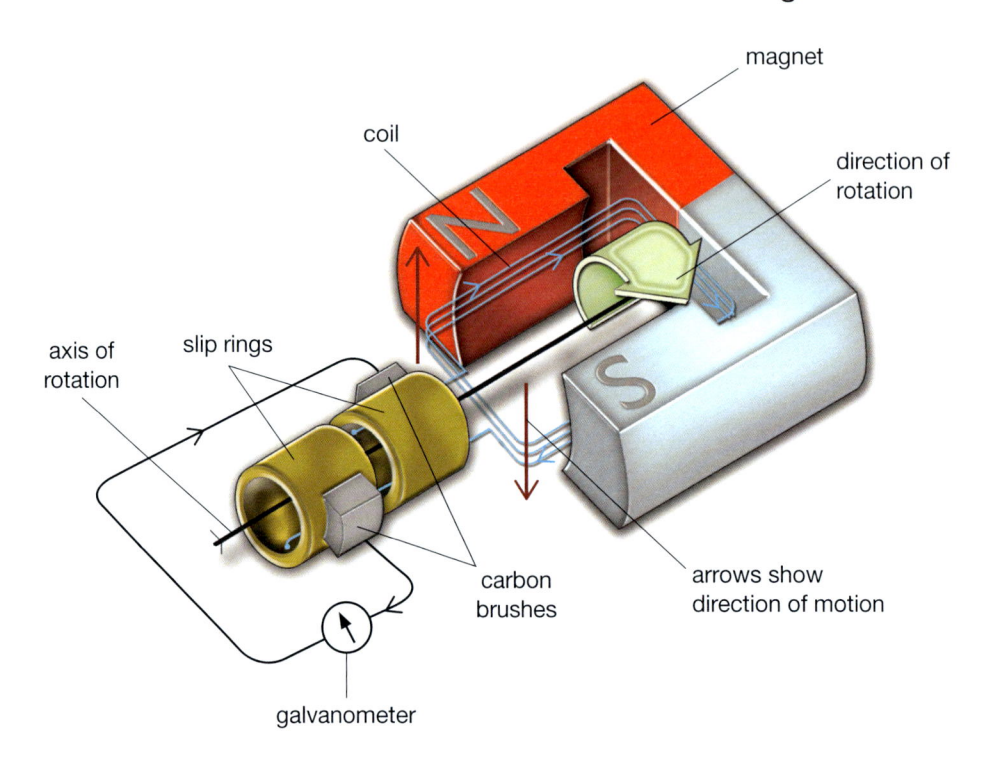

○ Generators often use a **rotating coil** between **fixed magnets**.

○ Each end of the coil is connected to a **slip ring** against which **carbon brushes** press. Carbon is a conductor.

○ The coil is made from copper wire as copper is a good electrical conductor.

○ The **galvanometer** is a sensitive ammeter that indicates the presence of a current.

❑ Again the electromagnetic force (e.m.f.) and current can be increased by:

- increasing the number of turns in the coil
- using a stronger magnet
- rotating the coil faster.

Unit 4 Electricity and magnetism   Notes

○ The diagram below shows a graph of voltage output against time, and how the output relates to the position of the coil in the magnetic field.

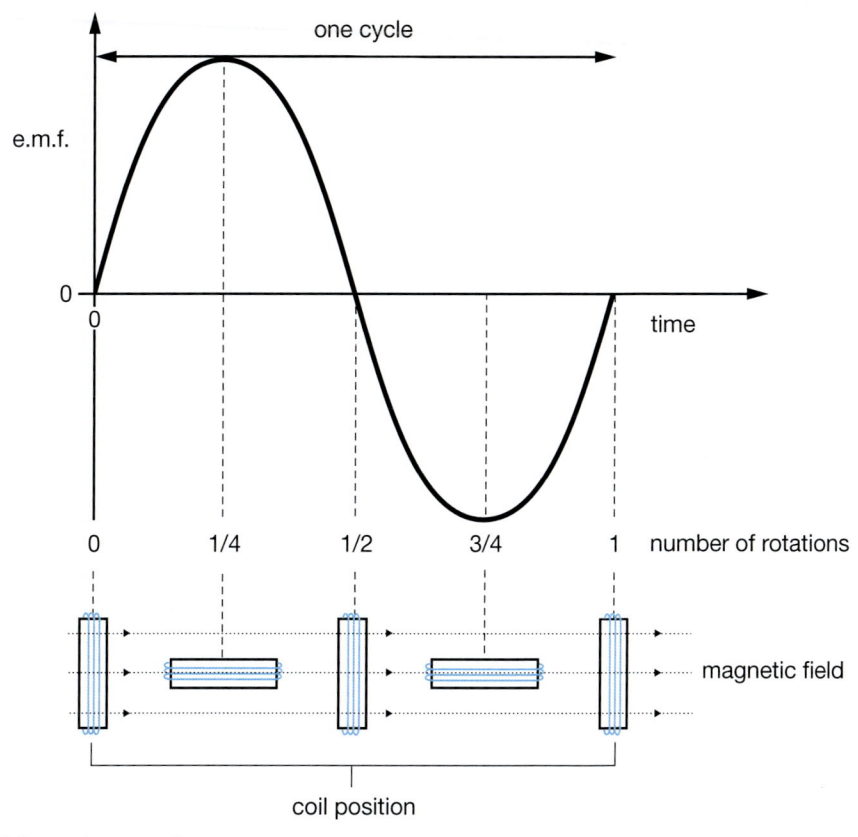

How it works

○ The coil is rotated in the magnetic field, **cutting** the magnetic field lines so that an alternating e.m.f. is induced.

○ This produces a.c. (alternating current) in an external circuit.

○ The direction of the current in each part of the coil can be found by applying Fleming's right hand rule. The current in the two sides of the coil is in opposite directions.

○ Because one side of the coil is always connected to the same slip ring, as the coil rotates the current in the circuit changes direction every half-cycle.

○ The e.m.f. and the current are at a **maximum** when the coil is **horizontal** as it is cutting the field lines at the greatest rate.

○ The e.m.f. and the current are **zero** when the coil is **vertical** as no field lines are being cut.

IGCSE Physics Summarised

Transformers

❏ A transformer is used to **change** an **alternating voltage** from one size into another.

❏ It is made by winding two insulated coils around a soft iron core, as shown below. These coils are known as the **primary** (**input**) and the **secondary** (**output**) coils. (There are, of course, many more turns on the coils of a real transformer.)

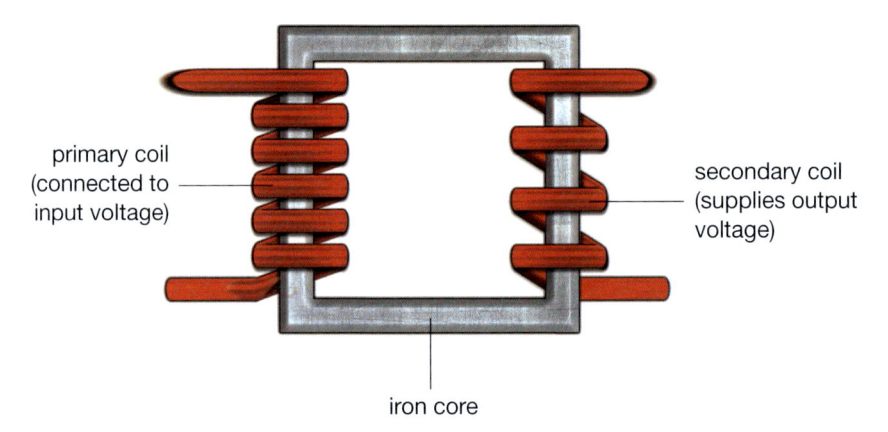

primary coil (connected to input voltage)

secondary coil (supplies output voltage)

iron core

❏ When an alternating current flows in the **primary** (input) coil, it sets up an **alternating magnetic field** in the primary coil and in the iron core.

○ This changing magnetic field is transferred through the core and into the **secondary** (output) coil.

○ This induces an **alternating e.m.f.** in the secondary (output) coil.

Top Tip

How transformers work in three easy steps:

Step 1 An **alternating current** in the primary coil sets up an **alternating magnetic field** in the primary coil.

Step 2 This alternating magnetic field is **transferred** through the iron core.

Step 3 The **alternating magnetic field** cuts through the **secondary coil** and **induces** an e.m.f.

Unit 4 Electricity and magnetism

Step-up and step-down transformers

❑ There are more **secondary turns** than primary turns in a step-up transformer.
The **secondary voltage** is larger than the primary voltage, i.e. it is stepped up.

step-up transformer

primary coil — secondary coil

❑ There are fewer **secondary turns** than primary turns in a step-down transformer. The **secondary voltage** is lower than the primary voltage, ie. it is stepped down.

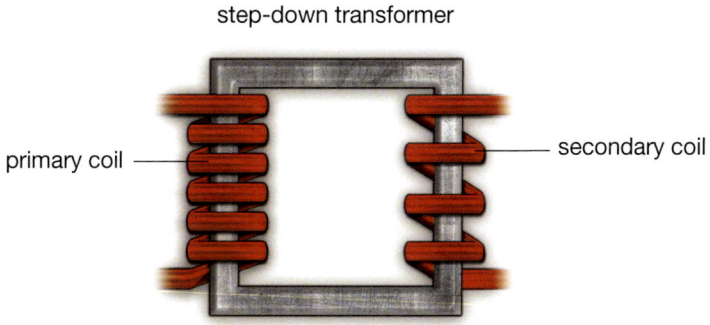

step-down transformer

primary coil — secondary coil

❑ The number of turns and voltage in the primary and secondary coil are related by the following expression:

$$\frac{V_p}{V_s} = \frac{N_p}{N_s}$$

V_s = secondary voltage (V)
V_p = primary voltage (V)
N_s = number of secondary turns
N_p = number of primary turns

❏ **Example**

A transformer transforms 240V a.c. to 12V a.c. for a model car racing set.
The secondary coil has 50 turns. Calculate the number of turns on the primary coil.

Step 1 List the information in symbol form and change into appropriate and consistent SI units if required.

$$V_p = 240V$$
$$V_s = 12V$$
$$N_s = 50$$
$$N_p = ?$$

Remember: There are no units for the number of turns.

Step 2 Use and rearrange the correct formula.

$$\frac{V_p}{V_s} = \frac{N_p}{N_s} \implies N_p = \frac{N_s \times V_p}{V_s}$$

Step 3 Calculate the answer by putting the numbers into the formula.

$$N_p = \frac{N_s \times V_p}{V_s} = \frac{50 \times 240}{12} = 1000$$

Efficiency of transformers

○ Transformers are **very efficient**; their power output is almost as high as their power input.

○ If a transformer could be made **100% efficient** then the power leaving the transformer would equal the power coming into the transformer.

$$P_p = P_s$$
$$I_p V_p = I_s V_s$$

P_p = primary power (W)
P_s = secondary power (W)
V_p = primary voltage (V)
I_p = primary current (A)
V_s = secondary voltage (V)
I_s = secondary current (A)

Unit 4 Electricity and magnetism   Notes

○ Note that if V_s is greater than V_p, as in a step-up transformer, then I_s must be less than I_p.

○ **No device is 100% efficient** and the energy output is always less than the energy input. Even though transformers are not 100% efficient, we assume that they are in calculations.

○ One reason why transformers are not 100% efficient is because the **resistance** in both the primary and secondary coils causes heating.

○ Their efficiency might also be affected by the primary coil's magnetic field not linking the secondary coil with maximum effect.

Transmission of electricity

❏ Power supplied to our home is generated in a power plant.

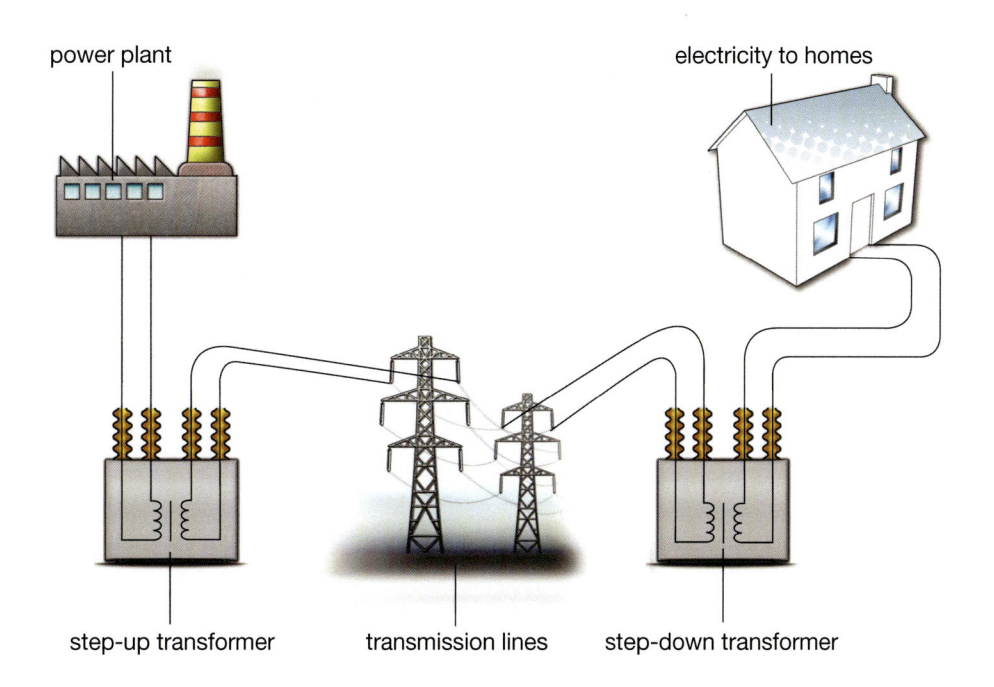

step-up transformer   transmission lines   step-down transformer

❏ **Step-up** transformers are used to **increase the voltage** and to **decrease the current** in cables used in transmission lines.

225

○ The current is decreased **deliberately** in order to reduce the power 'loss' in the cables (when current flows through the cables they heat up and energy is wasted).

The power 'loss' in the cables can be calculated using the following formula. Note that a smaller current will give a smaller power 'loss'.

$$P = IV = I(IR) = I^2R$$

P = power (W)
I = current (A)
R = resistance (Ω)

❏ So, by using step-up transformers to reduce the current, the **heating effect** is **minimised**. This is an advantage as it means cheaper and thinner cables can be used in transmission lines.

❏ **Step-down** transformers are used to **reduce the voltage** and increase the current before the electricity enters our homes.

❏ Transformers **only** work using **a.c.** and so that is why the current entering our homes is **a.c.** rather than **d.c.**

Note

Remember: Transformers will only work if there is an **alternating magnetic field**. This alternating magnetic field is only present if there is an **alternating current**. Therefore transformers will only work using a.c. and not d.c.

Unit 5 Atomic physics

Section 5.1 The nuclear atom

Atomic model

Rutherford's scattering experiment

- In 1910 Ernest Rutherford devised an experiment to investigate atomic structure.

- Rutherford's experiment confirmed that the atom is made up of a **very dense nucleus** containing most of the mass and that the rest of the atom is mainly **empty space**.

- This experiment was carried out by Hans Geiger and Ernest Marsden and involved firing alpha (α-) particles from a radioactive source at a piece of very thin gold foil in a vacuum.

- A zinc sulfide screen was used as a detector. The α-particles were detected by the **scintillations**, or flashes of light, that they caused on the screen.

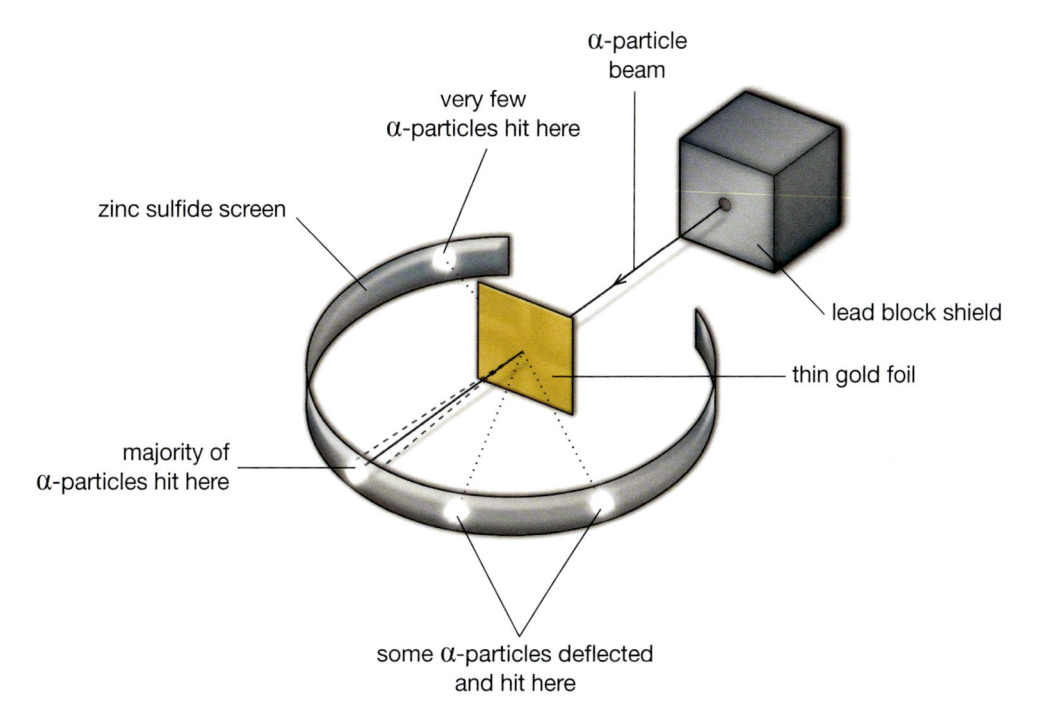

○ The results of this experiment were:

- The majority of the α-radiation passed straight through the foil.
- Some particles were deviated through fairly large angles of up to 90°.
- Very few were deflected at angles greater than 90° (i.e. they bounced back).

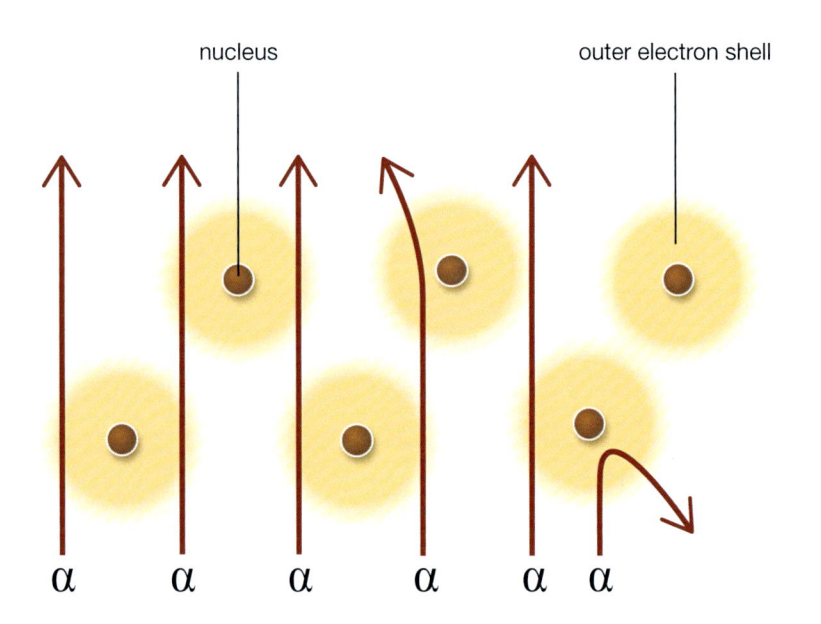

○ Rutherford concluded that:

- Most of the atom must be made up of empty space because most of the α-radiation passed straight through unaffected.
- The atom has a concentrated positive charge in the centre, because some of the positively charged α-particles were deflected by an appreciable angle or even bounced back.
- Since very few α-particles bounced straight back, this region of positive charge that repels the positive α-radiation must be very small in size and very dense.

He named this very dense and small centre the **nucleus**.

❏ We now know that the nucleus of an atom contains positive particles called **protons** and neutral (uncharged) particles called **neutrons**. The 'empty space' of the atom around the nucleus contains tiny negatively charged **electrons** in orbits or shells.

Unit 5 Atomic physics

Nucleus

- [ ] The number of **protons** in the nucleus is known as the **proton number Z** or **atomic umber**. In a **neutral** (not ionised), atom, the atomic number is also equal to the number of **electrons** orbiting the nucleus.

- [ ] The **nucleon number A** or **mass number** is the number of nucleons (protons and neutrons) in the nucleus.

- [ ] This is often written in the following format, known as nuclide notation:
  $^A_Z X$ where **X** is the **chemical symbol** for the element.

- [ ] Each **element** has a different **atomic number** and the elements are arranged in order of **increasing** atomic number in the periodic table.

- [ ] The masses and charges of the subatomic particles are given in this table in terms of the mass and charge of the proton.

Particle	Relative mass	Relative charge	Symbol
proton	1	+1	$^1_1 p$
neutron	1	0	$^1_0 n$
electron	1/2000	−1	$^0_{-1} e$

- [ ] This diagram is a simple representation of atomic structure.

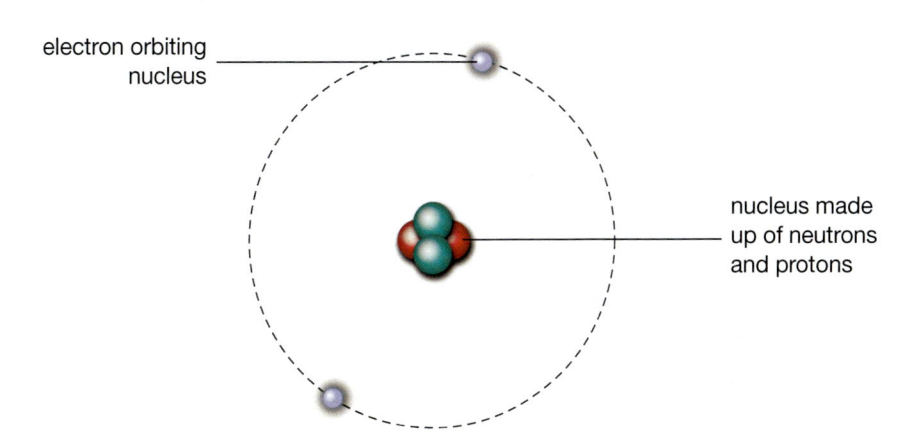

electron orbiting nucleus

nucleus made up of neutrons and protons

Isotopes

❑ Nuclei with the same **atomic number** can have different **mass numbers**.

For example:

$^{12}_{6}C$ $^{14}_{6}C$

These nuclei are called carbon-12 and carbon-14. They are different **isotopes** of carbon.

❑ Isotopes of an element have the same number of **protons** but a different number of **neutrons** (i.e. same atomic number but different mass number).

○ The isotope carbon-14 is radioactive and can be used in **carbon dating** (see page 237).

Fission and fusion

○ Nuclear power reactors use a nuclear reaction called **nuclear fission**. Fission means splitting.

○ Two isotopes commonly used as nuclear fuels are uranium-235 and plutonium-239. Both these isotopes are large nuclei and can be relatively easily split, especially when neutrons collide with them.

○ **Nuclear fusion** is the process that occurs in the Sun and other stars to create energy. In simple terms, hydrogen nuclei combine to form helium nuclei and release energy.

Section 5.2 Radioactivity

Detection of radioactivity

- [ ] Many nuclei of elements in the periodic table are **stable** but some are **unstable**. Unstable nuclei undergo **radioactive decay**.

- [ ] When a radioactive nucleus decays it may emit one or more of the following: **alpha particles**, **beta particles** or **gamma rays**.

- [ ] The radiations alpha, beta and gamma have their own individual characteristics, as shown in the table below.

	Alpha particle	Beta particle	Gamma ray
Nature	2 protons and 2 neutrons (helium nucleus)	electron	electromagnetic radiation
Symbol	α or ^4_2He	β or $^{\ \ 0}_{-1}e$	γ
Charge	positive	negative	uncharged
Affected by magnetic and electric fields	yes	yes	no
Penetrating power	weak – stopped by thin paper	moderate – stopped by a few mm of aluminium	strong – only stopped by many cm of lead or many m of concrete
Relative ionising effect	strongest	medium	weakest
Dangerous	yes	yes	yes

- [ ] There is always radiation present all around us. This is known as **background radiation**. Background radiation comes from **cosmic rays**, **rocks** and **atmospheric gases**. A small amount (about 3%) of background radiation comes from human-made sources such as medical equipment, nuclear power stations and nuclear weapons testing.

- ❏ **Ionising radiation** is powerful enough to knock electrons from atoms, leaving them positively charged. These positively charged particles are called positive **ions**. Alpha, beta and gamma radiation all have an ionising effect, to some degree (see the table on page 231).

- ❏ By ionising atoms, radiation may cause chemical reactions to occur.

- ❏ This is particularly dangerous if these reactions occur within the living cells of humans. Radiation can cause **sterility**, **anaemia**, **hair loss** and **cancer**, amongst other things.

- ❏ All radiation produced by radioactivity can be detected using **photographic film**, a **cloud chamber**, or more commonly a **Geiger–Müller tube** as shown below.

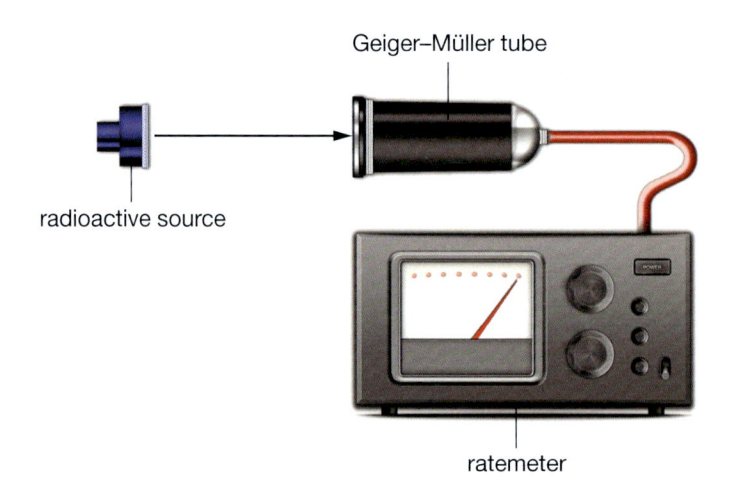

- ❏ α-particles are stopped by paper, β-particles are stopped by 3mm of aluminium, and γ-rays are greatly reduced in intensity by lead.

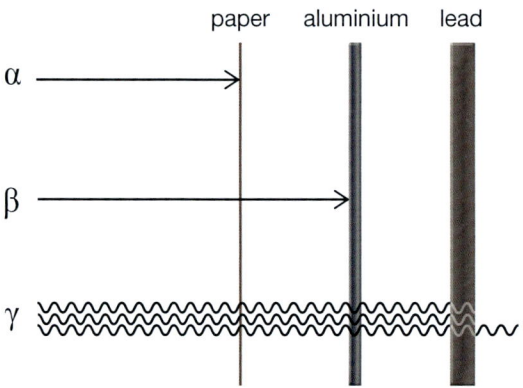

Unit 5 Atomic physics   Notes

❏ By using a Geiger–Müller tube and sheets of paper and aluminium, it is possible to deduce what type of radiation a radioactive source is emitting.

❏ **Example**

A student designs an experiment to investigate what type of radiation a radioisotope emits, as shown below. The ratemeter measures radiation in counts/minute.

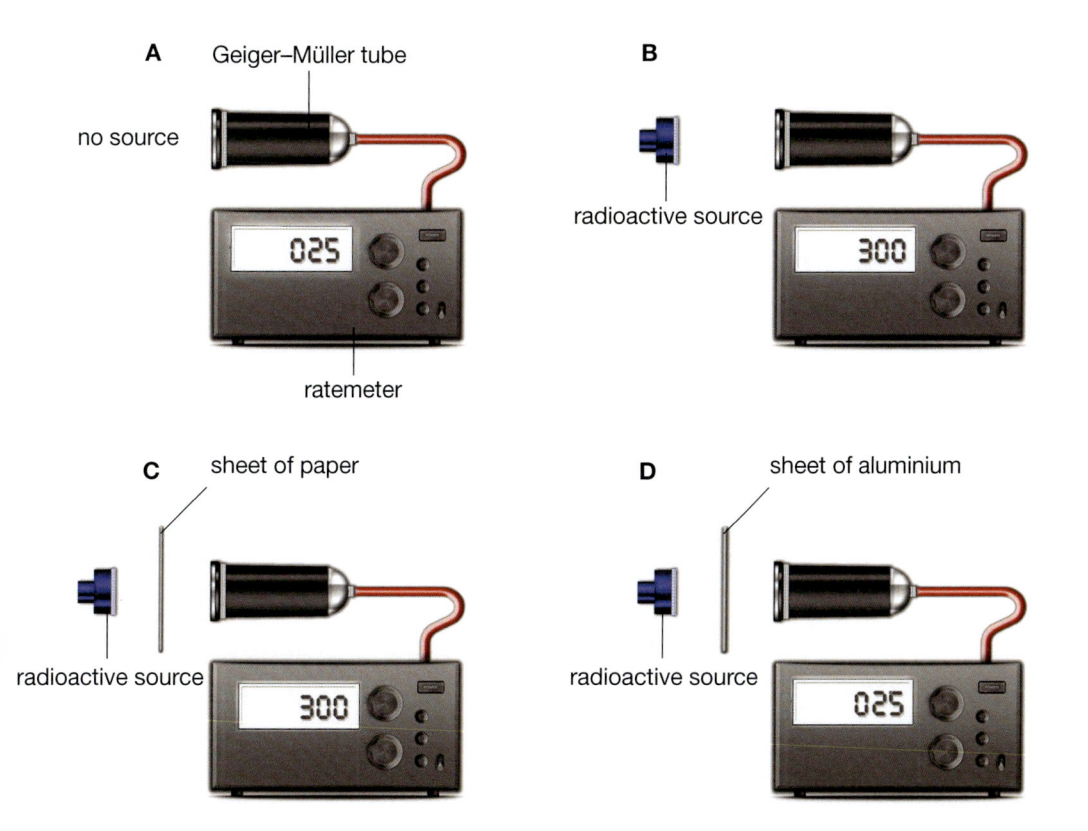

A Geiger–Müller tube
no source
025
ratemeter

B
radioactive source
300

C sheet of paper
radioactive source
300

D sheet of aluminium
radioactive source
025

The following measurements were taken by the student:

A   Detector and no source, giving a background reading of 25 counts/minute.

B   Detector and radioactive source, giving a reading of 300 counts/minute.

C   Detector, source and paper, giving a reading of 300 counts/minute.

D   Detector, source and aluminium, giving a reading of 25 counts/minute.

The student concludes that only β-particles are emitted by the radioisotope.
Explain how the results obtained show that only β-particles are emitted.

Answer

When paper was placed between the source and the detector in diagram **C**, the reading did not change so there can be no α-radiation emitted as this would have been stopped by paper.

When the aluminium was placed between the source and the detector in diagram **D**, the reading dropped to the background value, showing that the radiation was stopped by the aluminium. Therefore the radiation must be β-particles.

There can be no γ-rays present. If there were, they would not be stopped by the aluminium, so the reading would not drop in diagram **D** to the background value.

Characteristics of the three kinds of emission

- Radioactive emission is a completely **random** process. It is impossible to predict when a particular nucleus will decay.

- α-particles are the most ionising (refer back to the table on page 231), so α-radiation is potentially the most **harmful** if ingested into our bodies. However, α-particles are not very penetrating and so cannot get past our skin when produced outside the body. Consequently, α-radiation is more dangerous within our bodies than on the outside of our bodies.

- β-particles and γ-rays are **weaker** at ionising if ingested into our bodies but can penetrate our skin more easily when travelling through air. This is what makes them dangerous from outside the body.

- α-particles are **positively** charged and β-particles are **negatively** charged. Because they are charged, both α-particles and β-particles are affected by electric and magnetic fields.

Unit 5 Atomic physics

- ○ γ-rays have no charge and therefore **are not** influenced by an electric or magnetic field.

- ○ The diagram shows how the three kinds of radiation behave in an **electric field**.

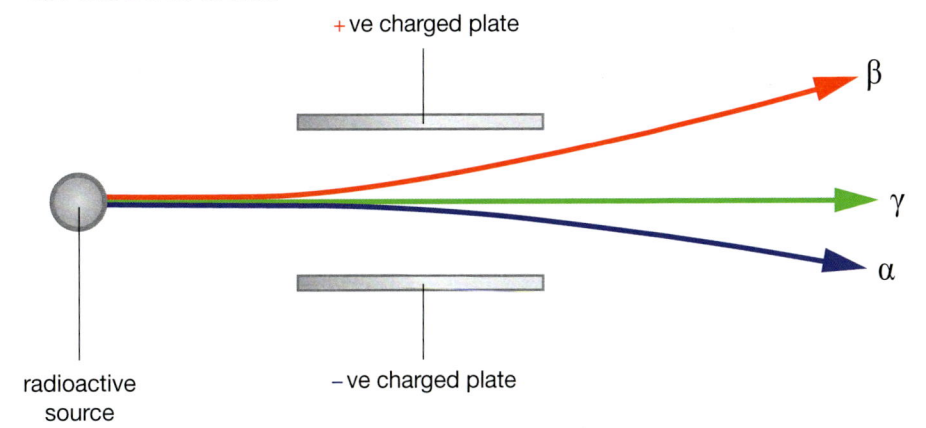

- ○ α-particles are **deflected less** than β-particles because they have a much **greater mass** and therefore need a much **bigger force** to deflect them.

- ○ The diagram below shows how the three kinds of radiation behave in a **magnetic field**.

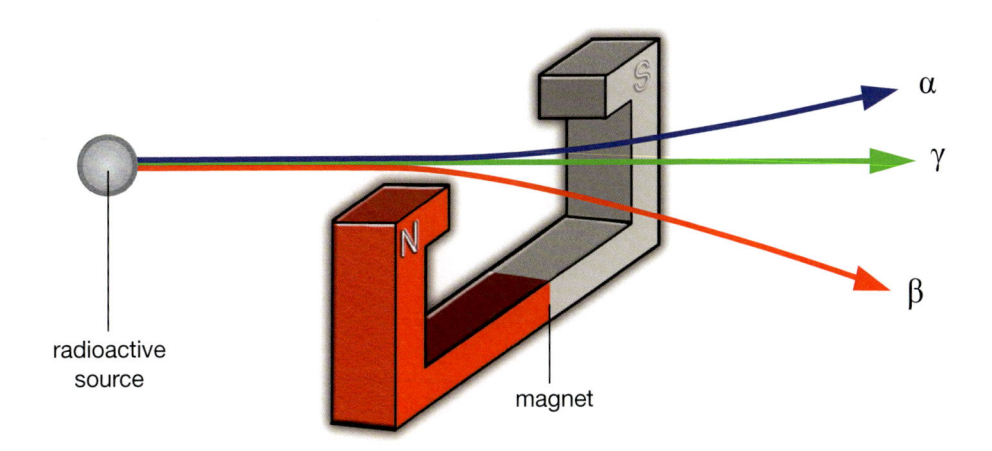

- ○ The deflection of α- and β-particles in a magnetic field can be found by using **Fleming's left-hand rule** (see page 213).

- ○ **Remember:** Fleming's left-hand rule is based on conventional current (positive to negative).

235

Uses of radioactive isotopes

○ γ-rays can be used to **sterilise medical equipment** because they kill bacteria.

○ β-particles can be used to monitor the **thickness of paper**. In this thickness-monitoring process the number of β-particles that pass through the paper is inversely related to the thickness of the material. If the paper becomes too thick, fewer β-particles are detected. If the paper becomes too thin, more β-particles are detected.

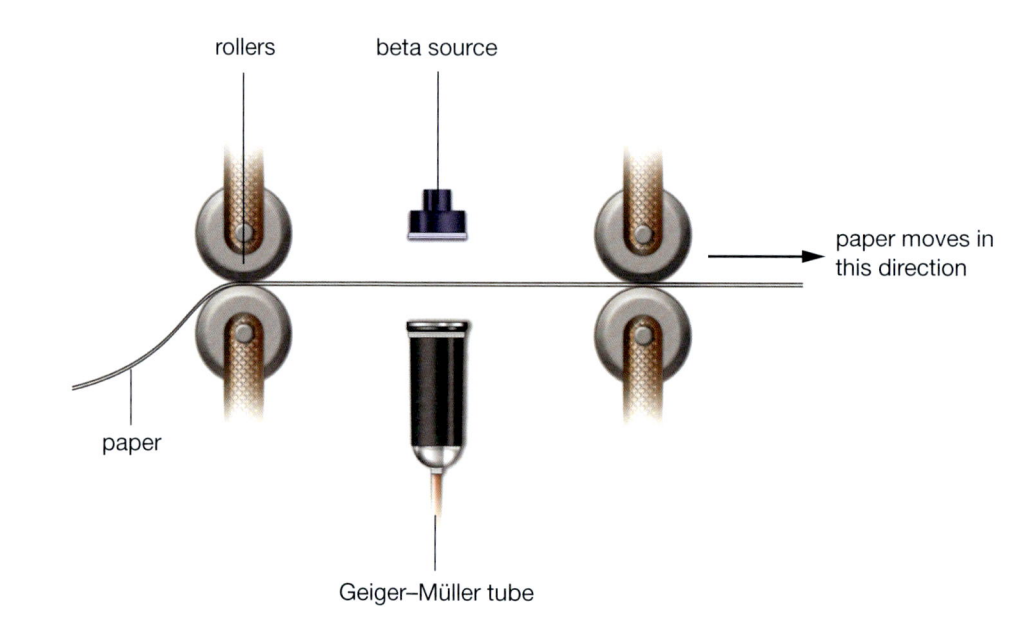

○ α-particles would not pass through at all and γ-radiation would pass through unimpeded. So neither α-particles nor γ-rays would be useful in this process.

○ **Medical tracers** are used to detect blockages in vital organs. A small amount of radioactive isotope is injected into a patient's bloodstream. Such isotopes usually emit γ-radiation, which will pass through the body to an external imager. The imager can follow the path the isotope takes. The isotopes have very short half-lives (see page 238), so they are quickly eliminated from the body.

○ γ-rays can be used in **radiotherapy**; beams of γ-radiation are fired directly at **cancer cells** to kill them.

Unit 5 Atomic physics

○ An isotope of carbon is used in **radioactive carbon dating**. All living organisms contain a small amount of carbon-14, which has a half-life of 5700 years. When the organism dies, the remaining carbon-14 decays slowly. The ratio of carbon-14 to the non-radioactive carbon-12 can be used to calculate when the organism was last living. Carbon dating is therefore very useful in archaeology.

Radioactive decay

❏ Unstable elements undergo the random process of radioactive decay. The nucleus of an unstable atom breaks up to form a different nucleus (i.e. a different element) and releases energy.

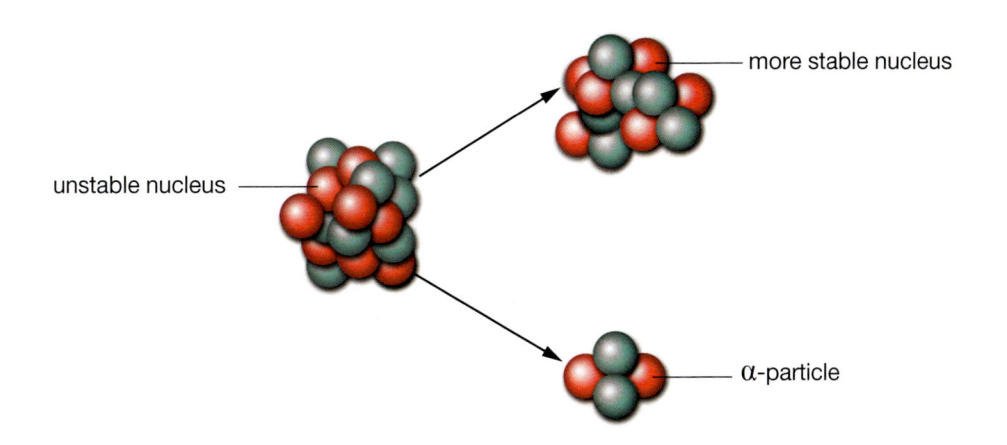

❏ An α-particle is identical to a helium nucleus, so it has an **atomic number of two** and a **mass number of four**.

○ In α-decay a nucleus loses **two protons** and **two neutrons**, so the mass number is reduced by four and the atomic number by two. For example, using nuclide notation:

$$^{226}_{88}\text{Ra} \rightarrow \,^{222}_{86}\text{Rn} + \,^{4}_{2}\alpha$$

The mass numbers and proton numbers balance on both sides of the equation:

226 = 222 + 4
88 = 86 + 2

- In β-decay a **neutron changes** into a **proton** and an **electron**. The **atomic number** of the nucleus therefore increases by **one** and the **mass number** stays the **same**. The new proton stays in the nucleus but the electron is expelled as a β-particle. For example, using nuclide notation:

$$^{131}_{53}I \rightarrow {}^{131}_{54}Xe + {}^{0}_{-1}\beta$$

Again the mass numbers and proton numbers balance on both sides of the equation.

- A β-particle is a high-energy (high-speed) **electron** emitted from the nucleus.

- γ-rays are emitted when a nucleus decays but is still in a slightly unstable state after emission of particles. γ-emission causes **no change** in atomic or mass number.

Half-life

- The **activity** of any radioactive source is measured in a unit called the **becquerel** (Bq).

- An activity of **1 Bq** is **one nucleus decaying per second**. If a source has an activity of 100 Bq, it follows that 100 radioactive nuclei are decaying per second.

- The activity of radioactive sources **decreases with time**.

- Some radioactive sources are **more unstable** than others and decay at a **faster rate**.

- **Remember:** 'Decay' does not mean that radioactive material disappears. The unstable nuclei of the material change to stable nuclei of a different element.

- The bigger the mass of a given source, the greater the activity.

- Radioactive decay is **not affected** by temperature or pressure; it is spontaneous.

Unit 5 Atomic physics

❑ Radioactive decay is a **random** process. There is no way to predict which nucleus in a radioactive substance will be the next to decay.

❑ However, the average time taken for **half** of the unstable nuclei in a sample of a particular radioactive isotope to decay is **always the same**. This time is known as the **half-life**.

❑ The half-life is the time taken for the **activity** of a radioactive isotope to **drop by half** of its original value.

❑ When calculating the half-life, the **count rate** must be corrected to account for **background radiation**. You must subtract the background count rate from the measured count rate before you start.

❑ The graph below shows a typical example of a radioactive isotope decaying over a period of time.

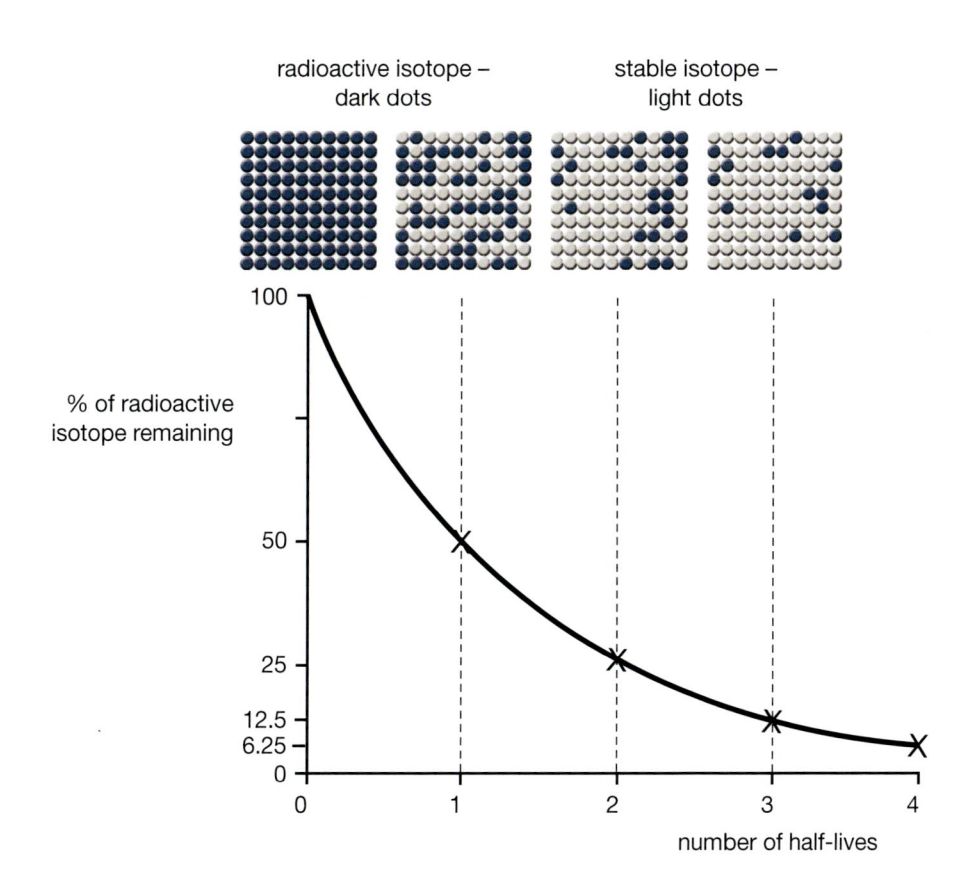

❑ Since the decay of an isotope is random, the curve is actually a curve of **best fit**.

IGCSE Physics Summarised

- ○ To calculate the half-life of an isotope without a graph, the count rate before and after and the total time elapsed are needed.

- ○ Although activity is measured in Bq (counts per second), counters are often calibrated in counts/min. It is acceptable to work in either unit, but never use both together in the same calculation.

- ❑ **Example**
  In an experiment, a radioactive isotope's activity falls from 200 Bq to 25 Bq in 75 minutes. Calculate its half-life.

  Answer
  200 Bq ➡ 100 Bq ➡ 50 Bq ➡ 25 Bq

  The activity has halved three times; therefore three half-lives have elapsed.

  half-life = 75 minutes ÷ 3
  half-life = 25 minutes

- ❑ To calculate the **activity** of an isotope after a period of time, the half-life, the starting activity, and the time elapsed are needed.

- ○ **Example**
  The half-life of a substance is 3.0 days. The initial count rate recorded next to a sample of this substance is 2050 counts/minute and the background radiation count rate is 50 counts/minute. Calculate the count rate 9.0 days later.

  Answer
  The actual initial count rate is 2000 counts/minute because background is 50 counts/minute.

The half-life is 3.0 days; therefore after 9.0 days three half-lives have elapsed.

2000 counts/minute → 1000 counts/minute → 500 counts/minute → 250 counts/minute

Therefore the activity is 250 counts/minute 9.0 days later, but the background radiation count rate of 50 counts/minute is constant, so the actual count rate is 300 counts/minute.

Safety precautions

- **Radioactive radiation** can **damage** living cells.

- α-particles, due to their strong ability to **ionise** other particles, are particularly dangerous to human tissue when inside the human body.

- γ-radiation outside the body is highly dangerous because of its high **penetrating** power.

- **Safety precautions** for storing and handling include, but are not limited to:
    - using forceps or robotic manipulators to hold radioactive sources
    - storing radioactive materials in thick lead containers
    - reducing the amount of time a person is exposed to radiation
    - wearing lead-lined clothing and gloves
    - working behind a lead–glass shield
    - wearing a film badge that alerts the wearer to the possibility of dangerous levels of radiation.

Additional support material

Physical quantities and units
Supplement level in grey tint

Quantity	Symbol	Unit (usual unit bold)
General physics		
period	T	s
length	l, h ...	**m**, mm, cm, km
area	A	**m²**, cm²
volume	V	**m³**, cm³
distance	d	**m**, cm, km
time	t	**s**, min, h, ms
speed	u, v	**m/s**, km/h, cm/s
acceleration	a	**m/s²**
mass	m	**kg**, g, mg
weight	W	N
acceleration of free fall	g	**m/s²**
gravitational field strength	g	N/kg
momentum	p	kgm/s
impulse		Ns
density	ρ	kg/m³, g/cm³
force	F	N
moment of a force	M	Nm
energy	E	**J**, kJ, MJ
work done	W, E	**J**, kJ, MJ
power	P	**W**, kW, MW
pressure	p	N/m², **Pa**
atmospheric pressure		mmHg
Thermal physics		
pressure	p	N/m², **Pa**
volume	V	m³, cm³, mm³
energy	E	**J**, kJ, MJ

Additional support material

Quantity	Symbol	Unit (usual unit bold)
power	P	**W**, kW, MW
time	t	**s**, min, h, ms
current	I	**A**, mA
voltage	V	**V**, mV
temperature	θ, T	°C
thermal capacity	C	J/°C
latent heat	L	J
specific heat capacity	c	**J/(kg°C)**, J/(g°C)
specific latent heat	l	**J/kg**, J/g

Properties of waves, including light and sound

Quantity	Symbol	Unit
frequency	f	**Hz**, kHz
wavelength	λ	**m**, cm
focal length	f	**cm**
angle of incidence	i	**degree** (°)
angle of reflection	r	**degree** (°)
angle of refraction	r	**degree** (°)
critical angle	c	**degree** (°)
refractive index	n	no units
distance	d	**m**, cm, km
time	t	**s**, min, h, ms
speed	v	**m/s**, km/h, cm/s

Electricity and magnetism

Quantity	Symbol	Unit
time	t	**s**, min, h, ms
charge	Q	**C**
current	I	**A**, mA
voltage/potential difference	V	**V**, mV
resistance	R	Ω
e.m.f.	E	V
energy	E	**J**, kJ, MJ
power	P	**W**, kW, MW

Atomic physics

Quantity	Symbol	Unit
half-life		**s**, min, h, year

How to use formulae effectively

- **Learn all the formulae** presented on the following pages. You must know them all for the supplementary level and the ones marked with a square for the core level.

- Read each examination question carefully and follow the steps below.

  Step 1 List the values given in the examination question using **symbol**, **value** and **units** that represent each quantity, including the symbol for the answer you are being asked to find.

  For example $v = 10 \text{cm/s}$
  $d = 5.0 \text{m}$
  $t = ?$

  Step 2 Change all units to appropriate and consistent SI units if required.

  For example $v = 0.1 \text{m/s}$

  Step 3 Write down the correct formula.

  For example $d = vt$

  Step 4 Rearrange the formula so that the subject of the formula you are trying to find is on its own on the left-hand side.

  For example $t = \dfrac{d}{v}$

  Step 5 Write down the figures in the formula and calculate the answer, **remembering to include units for calculated quantities**. (You may get marks for your working on the paper; remember this is the only way you can communicate with an examiner.)

  For example $t = \dfrac{5.0}{0.1} = 50 \text{s}$

Additional support material

Formulae

General physics

The triangles provide a learning aid. They are not an alternative way of writing the formula.

- period of pendulum = $\dfrac{\text{total time}}{\text{number of swings}}$

    $T = \dfrac{t}{\text{number}}$

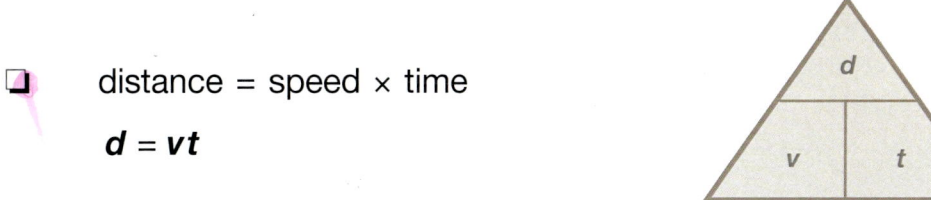

- distance = speed × time

    $d = vt$

- acceleration = $\dfrac{\text{final velocity} - \text{initial velocity}}{\text{time}}$

    $a = \dfrac{v-u}{t} = \dfrac{v}{t}$ if u is 0

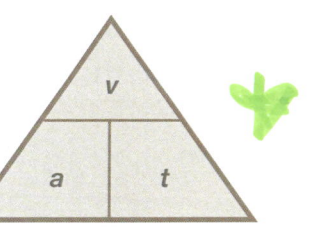

- weight = mass × gravitational field strength

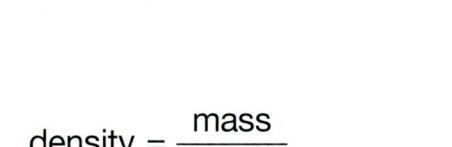

    $W = mg$

- density = $\dfrac{\text{mass}}{\text{volume}}$

    $\rho = \dfrac{m}{V}$

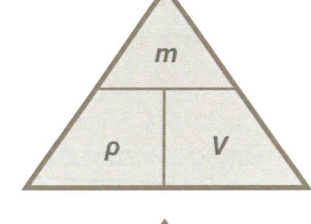

- force = mass × acceleration

    $F = ma$

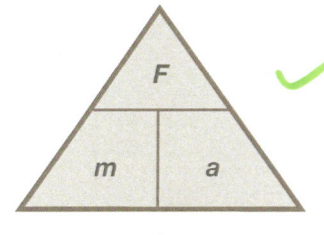

- force = spring constant × extension

    $F = kx$

Notes

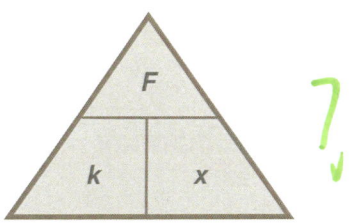

IGCSE Physics Summarised

❏ moment = force × perpendicular distance

$M = Fd$

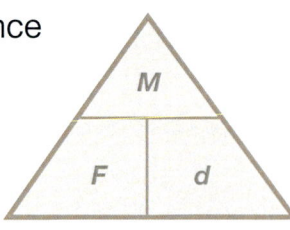

❏ sum of clockwise moments = sum of anti-clockwise moments

$F_1 d_1 = F_2 d_2$

○ momentum = mass × velocity

$p = mv$

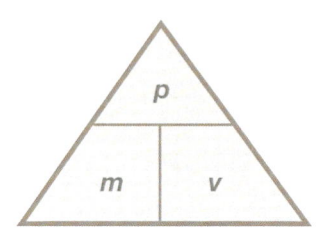

○ impulse = change in momentum

$Ft = mv - mu$

○ kinetic energy = $\frac{1}{2}$ × mass × velocity²

$k.e. = \frac{1}{2} mv^2$

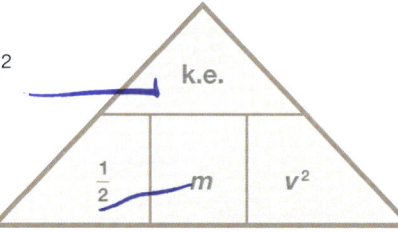

change in gravitational potential energy = mass × gravitational field strength × change in height

$g.p.e. = mg\Delta h$

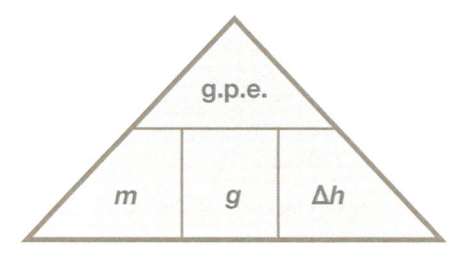

○ efficiency = $\dfrac{\text{useful energy out}}{\text{energy in}} \times 100\%$

efficiency = $\dfrac{E_{out}}{E_{in}} \times 100\%$

○ efficiency = $\dfrac{\text{useful power output}}{\text{power input}} \times 100\%$

efficiency = $\dfrac{P_{out}}{P_{in}} \times 100\%$

○ work done = force × distance

$W = Fd$

○ energy transferred = power × time

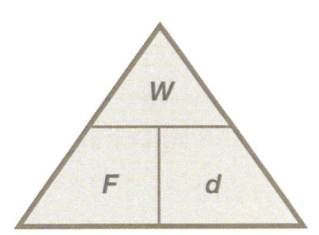

$\Delta E = Pt$

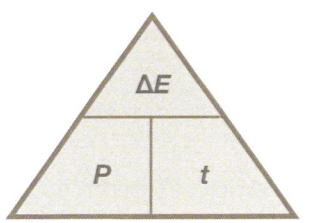

○ pressure = $\dfrac{\text{force}}{\text{area}}$

$p = \dfrac{F}{A}$

○ fluid pressure = density × gravitational field strength × height

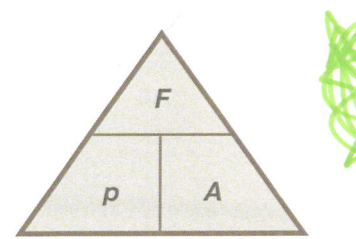

$p = \rho g h$

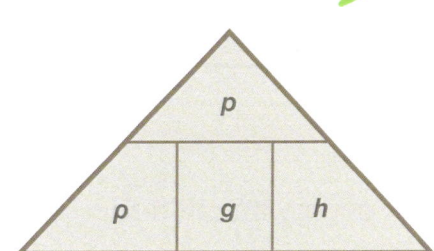

Thermal physics

- Boyle's Law:
  pressure × volume = constant
  pV = constant
  $p_1 V_1 = p_2 V_2$

- thermal capacity = mass × specific heat capacity
  $C = mc$

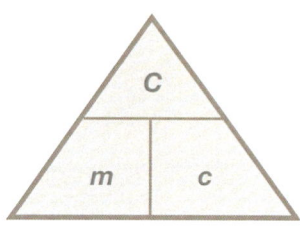

- change in energy = mass × specific heat capacity × change in temperature
  $E = mc\Delta\theta$

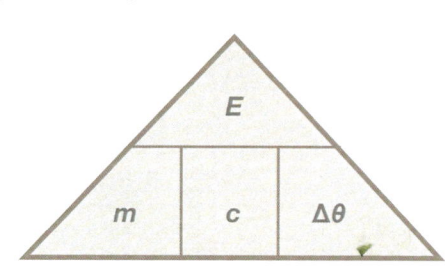

- energy transferred = mass × specific latent heat
  $E = ml$

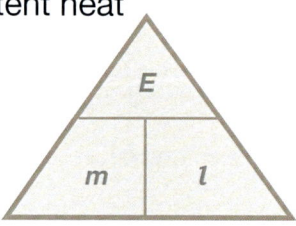

- energy transferred = power × time
  $\Delta E = Pt$

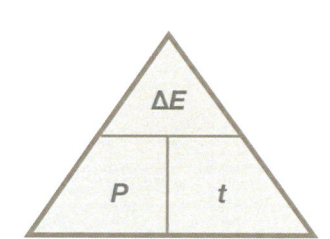

Additional support material

- power = current × voltage

  $P = IV$

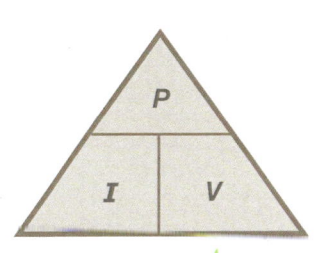

- energy transferred = current × voltage × time

  $E = IVt$

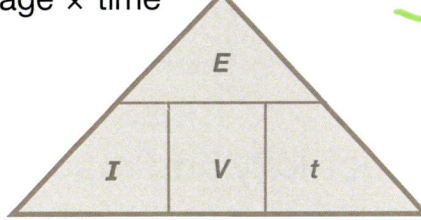

Properties of waves, including light and sound

- velocity = frequency × wavelength

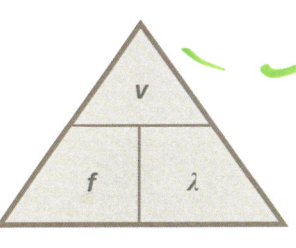

  $v = f\lambda$

- frequency = $\dfrac{1}{\text{period}}$

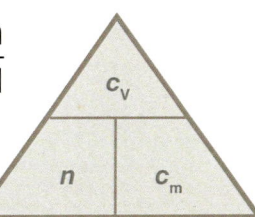

  $f = \dfrac{1}{T}$

- refractive index = $\dfrac{\text{sine of angle of incidence, } i}{\text{sine of angle of refraction, } r}$

  $n = \dfrac{\sin i}{\sin r}$

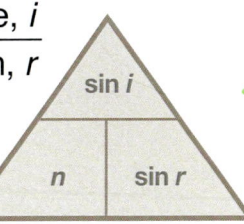

- refractive index = $\dfrac{\text{speed of light in vacuum}}{\text{speed of light in material}}$

  $n = \dfrac{c_v}{c_m}$

- refractive index = $\dfrac{1}{\text{sine of critical angle}}$

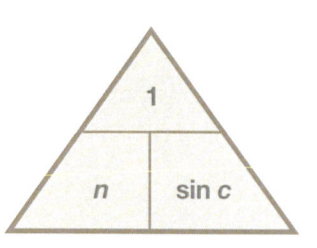

  $n = \dfrac{1}{\sin c}$

- distance = speed × time

  $d = vt$

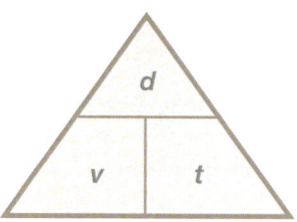

Electricity and magnetism

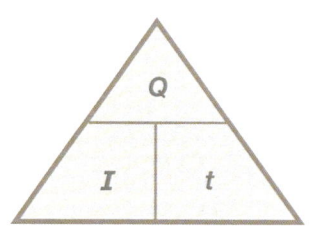

- current = $\dfrac{\text{charge}}{\text{time}}$

  $I = \dfrac{Q}{t}$

- resistance = $\dfrac{\text{voltage}}{\text{current}}$

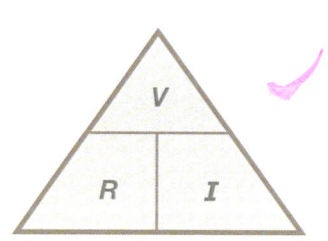

  $R = \dfrac{V}{I}$

- total resistance in series: $R_t = R_1 + R_2$

- total resistance in parallel: $\dfrac{1}{R_t} = \dfrac{1}{R_1} + \dfrac{1}{R_2}$

- power = current × voltage

  $P = IV$

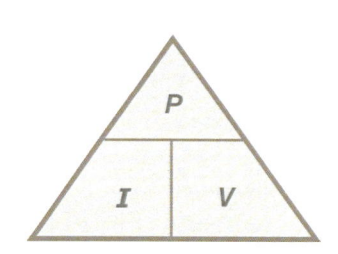

Additional support material

- energy transferred = power × time
  $\Delta E = Pt$

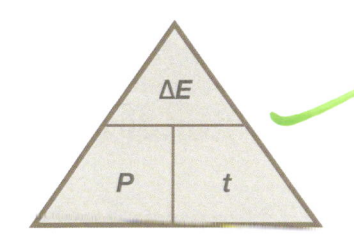

- power = current² × resistance
  $P = I^2R$

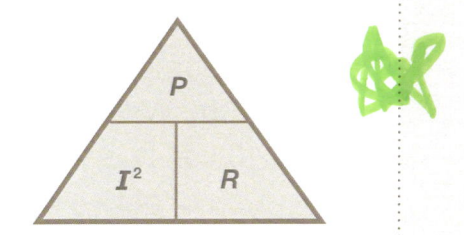

- energy transferred = current × voltage × time
  $E = IVt$

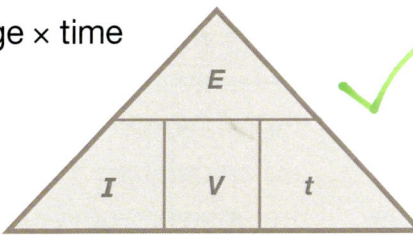

- For a transformer:

  $$\frac{\text{voltage in primary coil}}{\text{voltage in secondary coil}} = \frac{\text{number of turns in primary}}{\text{number of turns in secondary}}$$

  $$\frac{V_p}{V_s} = \frac{N_p}{N_s}$$

- For a 100% efficient transformer:

  voltage in primary × current in primary
  = voltage in secondary × current in secondary

  $V_p I_p = V_s I_s$

Working with numbers

Understanding significant figures

❏ Numbers can be expressed in many different ways. Let us say the calculated value for a particular quantity was 2. One student may write it as 2 and another as 2.0. They may appear the same but they mean different things. 2 has one significant figure. Writing 2 means the value is 1.5 or above but below 2.5. 2.0 has two significant figures. Writing 2.0 means the value is 1.95 or above but below 2.05.

❏ In physics calculations, many numbers may appear on the calculator, e.g. 5.046754327. Expressing your answer like this is **wrong** because it claims you know the answer far more precisely than any instrument you have used or any information you have been given. For example, let us say you are calculating the average time for an oscillation. If you use the answer above, you are claiming incorrectly that you can measure time with a stopwatch to this high degree of precision.

❏ As a general rule, giving answers to **two or three significant figures** is normally acceptable for the majority of Cambridge IGCSE Physics questions. It is usually possible to express values from a graph to **three significant figures**.

Rounding to three significant figures

❏ Rounding means reducing the digits in a number while trying to keep its value similar.

❏ In the number 6.0469812 the fourth figure is above 5, so you would increase the third figure by one unit, e.g. 6.05. This is **rounding up**.
In the number 6.0429812 the fourth figure is below 5, so you would leave the third figure as it is, e.g. 6.04. This is **rounding down**.

❏ **Remember:** Round up if the figure to the right is 5 or above and round down if it is below 5.

Additional support material

Standard form

☐ Many very large or very small numbers in physics are expressed in standard form, also commonly known as **scientific notation**.

☐ Standard form consists of two parts, a number between 1 and 10 followed by × 10 to the power of a number, known as the **index**.

e.g. 3.0×10^8 m/s is the speed of light

☐ The following numbers written out in full give:

$1.0 \times 10^5 = 100\,000$
$3.0 \times 10^8 = 300\,000\,000$
$3.56 \times 10^7 = 35\,600\,000$

The value of the index depends on how many places the decimal point has been moved. For example 35 600 000 becomes 3.56 by moving the decimal point 7 places to the left. The index is 7.

☐ To represent very small numbers, a negative index is used. The following numbers written out in full give:

$2.1 \times 10^{-5} = 0.000\,021$
$9.3 \times 10^{-8} = 0.000\,000\,093$
$3.56 \times 10^{-4} = 0.000\,356$

This time the decimal point moves to the right and the index is a negative number. For example 0.000 356 becomes 3.56 by moving the decimal point 4 places to the right. The index is −4.

☐ Values can be entered in standard form on a calculator by using the EXP button. For example:

$7.6 \times 10^4 =$ [7] [.] [6] [EXP] [4]

$3.2 \times 10^{-3} =$ [3] [.] [2] [EXP] [(−)] [3]

Do not include the 10. You would only use the 10 when using the [x^y] button, for example: [3] [.] [2] [×] [1] [0] [x^y] [(−)] [3] and so on.

❑ Sometimes, **prefixes** are used with SI units to simplify large or small values.

$$\begin{aligned} \text{kilo (k)} &= 10^3 & \text{milli (m)} &= 10^{-3} \\ \text{mega (M)} &= 10^6 & \text{micro (μ)} &= 10^{-6} \\ \text{giga (G)} &= 10^9 & \text{nano (n)} &= 10^{-9} \\ \text{tera (T)} &= 10^{12} & \text{pico (p)} &= 10^{-12} \end{aligned}$$

For example, you can write 2000 m as 2 km and 0.003 m as 3 mm.

Always change prefixed units to base units but notice that the kilogram is the base SI unit for mass.

Graphs

Plotting graphs

You may be advised which set of numbers is to be plotted on the *x*-axis and the *y*-axis.

❑ **Choosing the appropriate scale for the *x*- or *y*-axis**
- Find the maximum and minimum value needed for each axis. **Remember:** You do not have to start the axis from zero if the numbers given are all greater than zero unless you are trying to show that the two variables are proportional, in which case you must show that the line passes through the origin.
- Make the scale easy to interpret. Choose a scale where each square on the graph paper equals 1, 2, 5, 10, 50 or 100. For smaller values use 0.01, 0.02, 0.05, 0.1 or 0.5.

❑ **The axes**
- The *x*-axis will usually represent the **independent variable** and will rise in regular intervals, e.g. 2, 4, 6, etc. not 2, 4, 9 etc. The independent variable is the one you control. For example if you take a measurement every 10 s, time is your independent variable.

Additional support material

- The **y**-axis will usually represent the experimental results – the **dependent variable**. This will give rise to a straight line or a curve.
- Label the axes clearly with both variable and units.

❏ **Drawing the graph**
- The graph you draw (i.e. the points you plot) should cover as much of the graph paper as possible – three-quarters of the page is a good guide.
- Plot crosses (x) or encircled dots (⊙) rather than dots (·) on your graph. Re-check any points that do not appear to fit the pattern.
- Draw a smooth continuous line that will not necessarily pass through all the points, known as a **line of best fit**. If the graph looks like a straight line, then use a ruler.

○ **Finding the gradient of a line**
- Use $m = \dfrac{\Delta y}{\Delta x}$ to find the gradient of a line. Make sure you pick two points that are **on the line**. It is helpful to show the values you choose by drawing dotted lines from the axes (see below). Choose values as **far apart** as possible to give a more accurate gradient.

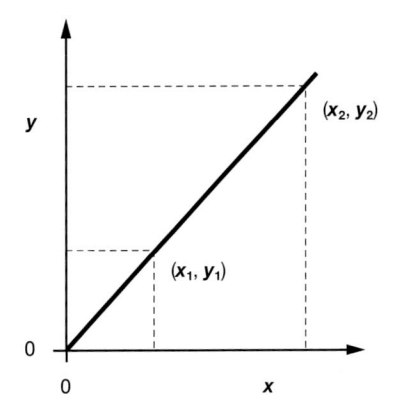

Example
Let (x_1, y_1) be (5.0, 50) and (x_2, y_2) be (10, 100).

$$m = \frac{y_2 - y_1}{x_2 - x_1} = \frac{100 - 50}{10 - 5.0} = 10$$

In many cases the gradient will have units. For example, if **y** is distance measured in metres and **x** is time measured in seconds, the gradient calculated above will be 10 m/s.

Understanding the graph you have plotted

❏ **Proportionality and linearity**

Many quantities in Physics are **directly proportional** to each other. Many formulae are derived from **straight-line (linear)** relationships.

The formula of a straight line on a graph is made up of a **y** term, an **x** term, and sometimes a number, and is written in the form of **y = mx + c**.

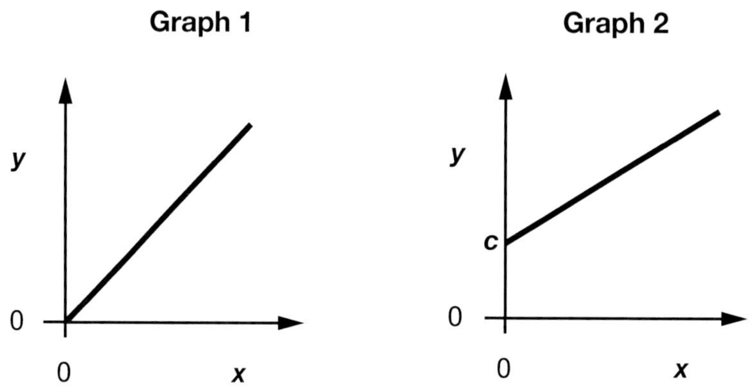

Graph 1

Graph 2

❏ *Graph 1 – directly proportional*
As **x** increases, **y** increases. The graph is a straight line and passes through the origin. Importantly the ratio of **x:y** is always the same. The graph formula is **y = mx**, where **m** is the steepness of the line, also known as the gradient:

$$m = \frac{y_2 - y_1}{x_2 - x_1}$$

❏ *Graph 2 – straight line with y-intercept*
As **x** increases, so does **y** as before. The difference is when **x = 0**, as this time **y** is not zero. The graph formula is **y = mx + c**, where **m** is the gradient. The **y**-intercept (the value of **y** when **x = 0**) is **c**.

Glossary

a.c. generator a device used to produce alternating current

acceleration how much an object's speed is increased per second – the rate of change of velocity of an object

acceleration due to gravity the acceleration of an object falling freely under gravity

accurate how close a measured value is to a known value

activity the rate of decay of nuclei in a radioactive sample

air resistance the frictional force on an object moving through air

alpha particle a type of nuclear radiation consisting of a helium nucleus ejected from an unstable nucleus

alternating current (a.c.) electric current that changes its direction repeatedly in a circuit; the charges flow one way and then the other way

ammeter an instrument for measuring electric current in amperes (A)

ampere (A) the base unit of current; it is the electric charge that flows past a point in one second

amplified increased in size

amplitude the maximum height or disturbance of a wave from its central equilibrium (rest) position

anaemia a condition in which a person does not have enough healthy blood cells to carry oxygen around the body

analogue a quantity that can be represented by a continuously varying signal

angle of incidence the angle measured between the normal and a ray of light arriving at a surface

angle of reflection the angle measured between the normal and a ray of light reflecting at a surface

angle of refraction the angle measured between a refracted ray and the normal to a surface

anode a positive electrode; another name for the positive terminal of a battery

atomic number see proton number

average speed the speed calculated by dividing total distance by total time taken

background radiation the radiation in the surrounding environment that we are exposed to all the time

balanced equal in size but opposite in sign or direction and therefore adding to zero

barometer an instrument used to measure atmospheric pressure

beta particle a type of nuclear radiation consisting of a high-speed electron emitted from an unstable nucleus

boiling point the temperature at which a liquid changes to a gas at normal pressure

Boyle's Law for a given mass of a gas at constant temperature, the volume of the gas is inversely proportional to the pressure

Brownian motion the motion of small particles in a fluid (liquid or gas), caused by collisions with molecules

calibrate to mark an instrument with a standard scale of readings such as the Celsius scale on a thermometer

cathode a negative electrode; another name for the negative terminal of a battery

cathode ray a beam of electrons travelling from a cathode (–ve) to an anode (+ve) in a vacuum tube

centre of mass the point in an object where all the mass appears to be concentrated; consequently also the point where all the weight appears to act

centripetal force the radial force required to keep an object moving in a circular path; the force always acts towards the centre of the circle

charge a property of certain materials, which can be positive or negative; a charged particle will experience a force if placed in an electric field

circuit breaker a safety device that switches off automatically when the current in it becomes too high; it can easily be reset

cloud chamber a device used to detect ionising radiation using alcohol vapour in an enclosed environment

collision where two or more objects strike each other and each object exerts a force on the other(s)

component part of a mechanical or electrical system

compression a region of a sound wave in which the particles are pushed close together

conduction the process by which thermal energy or electrical energy is transferred

conductor a material that transfers heat or allows charges to pass through it easily

conserved maintained at a constant overall; does not change for example, energy into and out of a system

constriction the act of narrowing, such as the narrowing of a tube

consumer a person who purchases goods and services such electrical energy for personal use

conventional current the imagined flow of electric charge, from the positive terminal in a battery round the circuit to the negative terminal

count rate the number of decaying radioactive nuclei detected per second or minute

critical angle the angle above which total internal reflection occurs

current the rate of flow of electric charge in a circuit

decay see radioactive decay

deceleration (also known as negative acceleration) how much an object's speed is decreased per second; the rate of change of velocity of an object

density the mass per unit volume of a substance

depleted used up to the point of it running out, such as an empty petrol tank in a car

depth the distance from the top to the bottom, such as in a swimming pool

determine find a quantity as a result of calculation or from a graph; often used when the quantity cannot be measured directly

deviated moved away from the original intended course

diffraction the spreading out of a wave when it passes through a gap or past an edge

digital signal a signal that has only two possible values and is represented by a 1 for on and a 0 for off

diminished smaller; for example, a diminished image is smaller than the object

diode a circuit component that allows current in one direction only

direct current (d.c.) electric current that is always in the same direction

directly proportional two quantities are directly proportional if their ratio is constant; as one increases, the other increases by the same percentage; if a graph of one quantity (y) is plotted against the other (x), the graph is a straight line that passes through the origin (0,0)

dispersion the splitting of white light into its component wavelengths (colours) due to refraction; for example, when white light falls onto a triangular prism

dissipate to disperse or scatter, to give out

domains can be thought of as small atomic magnets that need to line up for a magnetic material to become magnetised

drag a type of friction (sometimes called air resistance) that opposes the motion of an object

earthing when a charged object is connected to earth (or ground) and the charges flow to earth; for example, the casing of an electrical appliance is connected to the earth wire for safety

echo the reflection of sound from a surface heard some time after the original sound

efficiency a fractional measure (usually expressed as a percentage) of how effectively energy or power is transferred into a useful form in comparison to the total energy or power

elastic collision a collision in which the total kinetic energy is the same before and after the collision

electric current see current

electric field a region in space in which an electric charge will experience a force

electromagnet a coil of wire with an iron core that becomes magnetic only when there is an electric current in the coil

electromagnetic induction a method of producing an e.m.f. by moving a magnet into a coil of wire

electromagnetic (e.m.) spectrum the family of e.m. waves ranging from radio to gamma

electromagnetic (e.m.) waves energy travelling in the form of waves, which require no medium in which to travel

electromotive force (e.m.f.) the energy transferred to each coulomb of charge by a source of electrical energy such as a battery or power supply

electron a very small subatomic particle that is negatively charged and that exists in orbitals round the nucleus of an atom

electrostatic charges +ve or −ve charges that can be present on an insulator; insulators do not allow charges to move so they are static (**stay still**)

electrostatic induction a method of giving an object an electric charge without making physical contact with another charged object

endoscope a fibre optic device used to image the inside of living bodies

equilibrium when there is no net moment or no net force acting on a body

evaporation the process by which a liquid changes to gas or vapour below its boiling point

exert to make a physical effort such as when applying a force

extension the increase in the length of a spring when a load is attached

fluid a material that flows, such as any liquid or gas

focal length the distance from the centre of the lens to its principal focus

fossil fuel a source of energy such as coal, oil and gas formed from the remains of dead plants and animals over millions of years

frayed cables cables that are unravelled and have worn insulation

free fall an object falling under the influence of gravity alone

frequency the number of waves or vibrations passing a point per second

friction the force that opposes motion when two surfaces rub together

fuse a component designed to melt when a specified current value is exceeded, thus breaking the circuit

gamma ray highly penetrating electromagnetic radiation produced when an unstable atom decays

geothermal energy heat energy produced by nuclear processes in the Earth's core

gradient the ratio of the change in the quantity plotted on the y-axis to the corresponding change in the quantity plotted on the x-axis; here quantity plotted means, for example, distance /m rather than just distance

gravitational field strength the force in newtons exerted per kilogram of mass by gravity; at the Earth's surface this is approximately 10 N/kg

gravitational potential energy (g.p.e.) the energy possessed by an object due to its relative position; for example, its height above the Earth's surface

half-life the average time taken for half the nuclei in a radioactive sample to decay and form a new element

Hooke's Law the extension of an object is proportional to the load producing it, provided that the limit of proportionality is not exceeded

hydroelectric energy electrical energy produced using the gravitational potential energy of water stored in reservoirs in mountainous regions to turn turbines and drive generators

image optical reproduction of an object using lenses or mirrors (see real and virtual)

impulse the impulse acting on a body is equal to the product of the force acting on the body and the time for which it acts

incident ray a ray of light arriving at or striking a surface

induce to give rise to or cause something to happen by induction

induction see electromagnetic and electrostatic induction

inelastic collision a collision in which some of the kinetic energy is transferred into another form such as sound and heat

inertia the tendency of an object to resist any change in its motion, whether at rest or in uniform motion

infra-red radiation the portion of the electromagnetic spectrum between microwaves and visible light that is sometimes known as heat radiation

infrasound low-frequency sound, lower than 20 Hz (the normal limit of human hearing)

insulator a material that does not transfer heat or allow charges to pass through it easily

interact to act in a manner that causes objects to have an effect on each other

interaction see interact

internal energy the total kinetic and potential energy of the particles within an object

inversely proportional as one quantity increases, the other quantity decreases so that their product is constant

ionisation the process by which a particle (atom or molecule) becomes electrically charged by losing or gaining electrons

ionising radiation charged particles or high-energy light rays that ionise the material through which they travel

isotopes atoms of an element that have the same atomic number but different nucleon number (the same number of protons but a different number of neutrons)

kilogram (kg) the base unit of mass

kinetic energy (k.e.) the energy of an object due to its motion

lagging insulating material

lamina a flat object of constant thickness

laser a device that produces a concentrated narrow beam of single-frequency light

latent heat the energy needed to melt or boil a material without change of temperature

Lenz's Law the magnetic field generated by an induced current opposes the change in field that caused the current

light-dependent resistor (LDR) a resistor whose resistance varies according to the amount of light falling on it

light-emitting diode (LED) a diode that glows when current passes through it

limit of proportionality the point beyond which the extension no longer proportional to the load

linear in a straight line

load a force that causes a spring to extend

logic gate an electronic component whose output voltage depends on the input voltage(s)

longitudinal wave the particles of the medium through which wave travels move backwards and forwards along the same li as the direction of travel

loudness how we perceive the amplitude of a sound wave

luminous gives out light

magnetic field a region in space around a magnet or electric current in which a magnet will feel a force

magnetic material material that is attracted to or can be mac into a magnet; steel and iron are examples of magnetic mater

magnified enlarged; for example, a magnified image is larger than the object

magnitude the size of a quantity

manometer an instrument used to measure gas pressure

mass the amount of matter in an object; its base unit is the kilogram (kg)

mass number see nucleon number

matter an object that has a mass and occupies space

mean value an average value

medium a material that waves can travel through, such as ai water or glass

melting point the temperature at which a solid melts to beco a liquid at normal pressure

meniscus the curved surface of a liquid as seen, for example within a measuring cylinder

metre (m) the base unit of length

model a mental or physical representation that cannot be observed directly; usually used as an aid to understanding

moment the turning effect of a force about a pivot; the produ force and perpendicular distance from the pivot

momentum the product of mass and velocity of a body; it is measure of the quantity of motion in a body

monochromatic radiation of a single wavelength, such as laser light

negligible small enough to be ignored

neutral having no overall positive or negative charge

neutron an electrically neutral particle found in the nucleus of an atom

Newton's 1st Law an object will remain at rest or move at a steady speed in a straight line unless acted upon by an unbalanced force

Newton's 2nd Law an object will accelerate in the direction of an unbalanced force

normal a construction line drawn at right angles where a light ray strikes a surface such as a mirror or glass block

nuclear energy the energy stored in the nucleus of an atom

nuclear fission the process by which energy is released by the splitting of a large heavy nucleus into two or more lighter nuclei

nuclear fusion the process by which energy is released by the joining of two small light nuclei to form a new heavier nucleus

nucleon a particle such as a proton or neutron found in the nucleus of an atom

nucleon number (A) (also known as mass number) the number of protons and neutrons found in the nucleus of an atom

nucleus the very small and densely concentrated centre of an atom made up of protons and neutrons

nuclide a 'species' of a nucleus characterised by its atomic and mass numbers

Ohm's Law the current in a metal conductor is directly proportional to the potential difference across it provided the temperature remains the same

oscillation a repetitive motion, such as a pendulum swinging back and forth

parallax error the apparent change in position of a measurement caused by viewing the measurement at an angle less than or greater than 90 degrees to the scale

parallel circuit a circuit in which current has more than one path

penetrating passing through or into something

period the time taken for one complete cycle of oscillation, vibration or passage of a wave

periscope an optical device for viewing objects otherwise out of sight

perpendicular at right angles or 90 degrees

pitch how high or low a musical note sounds; is related to frequency: the higher the frequency the higher the pitch

pivot the fixed point about which a lever turns

potential difference (p.d.) the difference in electrical energy between two points of a circuit measured across the terminals of an electrical component such as a bulb; often referred to as voltage

potential divider usually consists of two or more resistors arranged in series across a power supply; designed to deliberately split the voltage in a circuit

power the rate at which energy is transferred or work is done

precise how close repeated measurements of the same quantity are to each other; precise measurements do not necessarily indicate accuracy

pressure the force acting per unit area at right angles to the surface

primary coil the input coil to a transformer

principal axis the line passing through the centre of a lens at right angles (perpendicular) to its surface

principal focus the point at which rays of light parallel to the principal axis converge after passing through a converging lens

propagation direction of travel

proportional when a change in one quantity is accompanied by a consistent change in the other

proton a positively charged particle found in the nucleus of an atom

proton number (Z) (also known as atomic number) the number of protons in the nucleus of an atom

radial moving or directed along the radius of a circle

radiation the transfer of energy by electromagnetic waves

radioactive decay the natural and random change of an unstable nucleus when it emits radiation

random a spontaneous event that cannot be predicted; for example, random decay means that it is not possible to predict when a particular nucleus will decay (but, statistically, it is possible to determine how many nuclei will decay in a given time)

range the difference between the minimum and maximum reading of an instrument; or the maximum distance a particle can travel

rarefaction a region of a sound wave in which the particles are further apart

ray a narrow beam of light

real image an optical image that can be formed on a screen

rectification the process of converting a.c. into d.c. with the use of one or more diodes

reflection the change in direction when a ray of light or sound wave bounces off a surface

refraction the change in direction of a light ray when passing from one material into another

refractive index the ratio of speed of light in a vacuum to that in a particular material; a material that has a large refractive index will refract light more than a material with a lower index

relay an electromagnetic switch

repulsion a force that causes objects to separate and move in opposite directions

resistance a measure of how difficult it is for current to pass through a circuit or part of a circuit; measured in ohms (Ω)

resistor a component in an electrical circuit that resists current

resultant force the net force acting on a body when two or more forces are unbalanced; effectively the replacement of all forces acting on an object with one equivalent force

ripple a small, uniform wave on the surface of water

scalar quantity a quantity with magnitude (size) only

scintillations flashes of light

second (s) the base unit of time

secondary coil the output coil of a transformer

sensitivity response to change; a sensitive instrument gives a large reading for a small change in the quantity being measured

series circuit a circuit in which current has one path

short circuit a low-resistance connection between two points, causing a large current to flow

SI an internationally agreed system of units based on the metric system

slip rings used to allow the passage of current to and from a coil in an a.c. generator

Snell's Law the ratio of sin i to sin r is a constant and is equal to the refractive index of the second medium with respect to the first

soft magnetic materials materials that, once magnetised, can easily be demagnetised

solenoid a coil of wire that becomes magnetised when a current is present in the coil

sound wave a longitudinal wave that carries sound energy from place to place

specific heat capacity (s.h.c.) the energy needed to raise the temperature of one kilogram of a substance by one degree Celsius

specific latent heat the energy needed to melt or boil one kilogram of a substance without a change in temperature

spectrum colours of light separated out in the order of their wavelengths

speed the distance travelled by an object per second

speed of light in a vacuum is 3×10^8 m/s

split ring used to allow the passage of current to and from a coil in a d.c. motor; also reverses the direction of the current every half turn

static electricity electric charge held by a charged insulator

sterilise to kill bacteria and clean

sterility the inability to have children

streamlined smoothed and somewhat rounded in order to reduce air resistance

temperature a measure of how hot a body is

terminal velocity the maximum velocity reached by an object moving through a fluid such as air; reached when the force due to its weight is equal to the force due to air resistance

thermal energy the internal energy a body has because of the motion of its particles

thermal expansion the expansion of a material due to a rise in temperature

thermistor a component whose resistance decreases with an increase in temperature and vice versa

thermocouple a thermometer made from two metal wires joined at the ends to form junctions

total internal reflection all light is reflected back from a surfac between materials; this happens as a result of the angle of incidence being greater than the critical angle

transducer any device or component that converts one form c energy into another

transformer a device used to change the voltage of an a.c. electricity supply

transmission lines power cables used to carry electricity from power stations to consumers

transverse wave a wave in which the vibrations or oscillations are at right angles to the direction of travel

truth table a summary table of the action of logic gate(s)

turbine a machine similar to a fan with blades that rotate when air, steam or water passes through; often used to generate electricity

ultrasound sound waves with frequencies greater than 20 000 Hz, which cannot be heard by the human ear

ultraviolet radiation the portion of the electromagnetic spectrum between visible light and X-rays, which can cause tanning of the skin

uniform constant

variable resistor a component whose resistance can be manually altered

vector a quantity with both magnitude (size) and direction

velocity the speed of an object in a particular direction

virtual image an image that cannot be formed on a screen

voltage a measure of the energy converted per unit charge passing through a component; also a measure of the amount of energy transferred to electrical form per unit charge by an electrical power supply, like a battery; measured in volts (V)

voltmeter a meter used for measuring the voltage (p.d) betwe two points

wavefront the set of points connected in space by a wave or vibration at the same instant; wavefronts generally form a continuous line or surface; for example, the lines formed by ripples on a pond

wavelength the distance between two adjacent identical poir on a wave

weight the downward force due to gravity acting on an object's mass

work done the product of force and distance in the direction the force

Glossary for examination terminology

This glossary (which is relevant only to science subjects) will prove helpful to candidates as a guide, but it is neither exhaustive nor definitive. Candidates should appreciate that the meaning of a term must depend, in part, on its context.

Define (the term(s) …) is intended literally, only a formal statement or equivalent paraphrase being required.

What do you understand by/What is meant by (the term(s) …) normally implies that a definition should be given, together with some relevant comment on the significance or context of the term(s) concerned, especially where two or more terms are included in the question. The amount of supplementary comment intended should be interpreted in the light of the indicated mark value.

State implies a concise answer with little or no supporting argument (e.g. a numerical answer that can readily be obtained 'by inspection').

List requires a number of points, generally each of one word, with no elaboration. Where a given number of points is specified this should not be exceeded.

Explain may imply reasoning or some reference to theory, depending on the context. It is another way of asking candidates to give reasons. The candidate needs to leave the examiner in no doubt why something happens.

Give a reason/Give reasons is another way of asking candidates to explain why something happens.

Describe requires the candidate to state in words (using diagrams where appropriate) the main points.

Describe and **explain** may be coupled, as may **state** and **explain**.

Discuss requires the candidate to give a critical account of the points involved.

Outline implies brevity (i.e. restricting the answer to giving essentials).

Predict implies that the candidate is expected to make a prediction not by recall but by making a logical connection between other pieces of information.

Deduce implies that the candidate is not expected to produce the required answer by recall but by making a logical connection between other pieces of information.

Suggest is used in two main contexts, i.e. either to imply that there is no unique answer (e.g. in physics there are several examples of energy resources from which electricity, or other useful forms of energy, may be obtained), or to imply that candidates are expected to apply their general knowledge of the subject to a 'novel' situation, one that may be formally 'not in the syllabus' – many data response and problem solving questions are of this type.

Find is a general term that may variously be interpreted as **calculate**, **measure**, **determine**, etc.

Calculate is used when a numerical answer is required. In general, working should be shown, especially where two or more steps are involved.

Measure implies that the quantity concerned can be directly obtained from a suitable measuring instrument (e.g. length using a rule, or mass using a balance).

Determine often implies that the quantity concerned cannot be measured directly but is obtained from a graph or by calculation.

Estimate implies a reasoned order of magnitude statement or calculation of the quantity concerned, making such simplifying assumptions as may be necessary about points of principle and about the values of quantities not otherwise included in the question.

Sketch, when applied to graph work, implies that the shape and/or position of the curve need only be qualitatively correct, but candidates should be aware that, depending on the context, some quantitative aspects may be looked for (e.g. passing through the origin, having an intercept). In diagrams, **sketch** implies that simple, freehand drawing is acceptable; nevertheless, care should be taken over proportions and the clear exposition of important details.

Index

a.c. generators 220–1
acceleration 10
   and forces 28–9, 33
   formulae 10, 13, 29
   of free fall (due to gravity) 18–19
   and speed–time graphs 9–11
activity 238–40
ADC (analogue to digital convertor) 201
air resistance 19, 20, 32
alpha (α) particles 231–5
   characteristics 231, 234–5
   dangers 241
   deflection in electric fields 235
   deflection in magnetic fields 235
   and radioactive decay 237
   scattering experiment 227–8
   stopping 232–4
alternating current (a.c.) 168, 220
ammeters 170
amplitude 125, 154
analogue signals 201
AND gates 202, 203
angle of incidence 131, 135
   measurement 138
angle of reflection 131
angle of refraction 135
   measurement 138
atmospheric pressure 79
atomic model (structure) 162, 227–9
atomic number (Z) 229

background radiation 231, 239
bar magnets 160
   magnetic field due to 156
barometers, mercury 79
becquerel (Bq) 238
beta (β) particles 231–5
   characteristics 231
   deflection in electric fields 235
   deflection in magnetic fields 235
   and radioactive decay 238
   stopping 232–4
bimetallic strips 95
binoculars 140
boiling 89, 110
Boyle's Law 90–3
Brownian motion 86
brushes, carbon 214, 220

Celsius scale 98
centres of mass 43–5
   irregularly shaped objects 44
   regularly shaped objects 43
   and stability 45
centripetal force 30
changes of state 110–13
charge
   electron 168
   unit and symbol 168
charges, electric see electric charges
charging
   by induction 166
   by rubbing 161, 162

chemical energy 53, 66, 67, 70
circuit breakers 206
circuit symbols 167
circuits, electronic see electronic circuits
circuits (electric) 186–200
   parallel 187–9, 191
   series 186–90
circular motion 30
collisions 61–5
compressions 149
condensation 110
conduction 118, 119
   applications 123, 124
   consequences 122
conductors, electrical 161
conservation of energy 53–4, 56–61
conservation of momentum 52
constant-volume gas thermometers 99
control systems 194–200
   input sensors 195
   output devices 195
   processors 195
   transducers 195–6
convection 118, 119, 120
   applications 123, 124
   consequences 122
converging (convex) lenses 141–4
critical angle 139
current 168–70
   alternating (a.c.) 168, 220
   direct (d.c.) 168
   measurement 170
   symbol and unit 168
current–voltage graphs
   filament lamps 180
   ohmic resistors 178

d.c. motors 214–15
deceleration, and speed–time graphs 9, 10
demagnetisation, methods of 159
density 22–7
   comparison of 22
   definition 22
   and floating 22
   gases 23
   irregularly shaped objects 26–7
   liquids 25–6
   determination of 23–7
   regularly shaped objects 23–4
   units 22
diffraction
   light 129
   waves 129–30
digital signals 201
diodes 196–7
direct current (d.c.) 168
direct proportionality 35, 256
dispersion, light 145
distance 6, 14
   from speed–time graphs 10, 12, 15
distance–time graphs 8

domains 157
double glazing 123
double insulated appliances 208

earthing 208
echoes 151–2
efficiency 72, 74
elastic collisions 61, 62–3
electric charges 161–6
   attraction and repulsion 163
   fields around 163–4
   static 161, 164–5
   test for 164, 165
electric circuits see circuits (electric)
electric fields 163–4
electric motors 181
electric plugs 208
electric relays 196, 199, 211
electrical conductors 161
electrical insulators 161
electricity
   generation 67–9
   hazards and safety measures 206–8
   transmission 225–6
electromagnetic induction
   coil 219
   wire 217–18
electromagnetic waves (spectrum) 127, 146–7
electromagnets 160, 211
electromotive force see e.m.f.
electronic circuits 194
electron(s) 161, 162, 228
   charge 168, 229
   mass 229
   symbol 229
electrostatic (static) charges 161, 164–5
electrostatics 161
e.m.f. 171
   induced 217–18
endoscopes 140
energy 53–72
   in collisions 61–5
   conservation of 53–4, 56–61
   forms of 53
   gravitational potential 55–9
   kinetic 54–9
   units 53
   useful 72
energy resources 66–72
energy transfer
   in electrical circuits 181
   see also heat (energy) transfer
equilibrium
   conditions for 39, 41–2
   stable 45
   unstable 45
evaporation 88–9
expansion, thermal see thermal expans

fibre optic cables 140
fission, nuclear 66, 230
Fleming's left-hand rule 213, 235
Fleming's right-hand rule 218
floor insulation 123
focal length 141
focal point 141
force(s)
  and acceleration 28–9, 33
  balanced 28
  centripetal 30
  on charged particles in magnetic field 216
  on current-carrying conductor 212–13
  due to gravity 18–20
  effects of 28
  friction 32, 54
  resultant 28–9, 31
  symbol and units 29
  turning effect of (moments of) 38–42
  see also weight
formulae 245–51
  effective use 244
  electricity and magnetism 250–1
  general physics 245–7
  light 249–50
  thermal physics 248–9
  wave properties 249–50
forward-bias 197
frequency 125, 153
friction 32, 54
fuse rating 206, 207
fuses 206
fusion, nuclear 67, 230

galvanometers 220
gamma (γ) rays 147, 231–5
  characteristics 231
  dangers 241
  path in electric fields 235
  path in magnetic fields 235
  and radioactive decay 238
  stopping 232–4
gases
  convection in 118, 119
  density 23
  pressure 79, 87, 90–3
  properties 84, 85
  temperature 87, 93
  thermal expansion 97
Geiger–Müller tube 232
geothermal energy 69, 70, 71
gold leaf electroscopes 165
graphs
  drawing 255
  gradients 8, 9, 11, 255
  linearity shown by 256
  plotting 254–5
  proportionality shown by 256
gravitational field strength 16–17, 19
gravitational potential energy 55–9
gravity, force due to 18–20

half-life 238–41
hard magnetic materials 160
heat 104

dissipation to surroundings 108–9
heat capacity see thermal capacity
heat (energy) transfer 118–24
heat loss, from homes 123
Hooke's Law 34–7
hydroelectric energy 69, 71

images
  describing 143–4
  real 141, 143
  virtual 132, 143
immersion heaters 181
impulse, and momentum 51
induced e.m.f. 217
induced magnetism 157
induction, charging by 166
inelastic collisions 61, 64–5
inertia 16
infra-red radiation (rays) 118, 147
insulation, in homes 123
insulators, electrical 161
internal energy 104
ionising radiation 232
ipods 181
isotopes 230

kinetic energy 54–9
kinetic molecular model of matter 83–93
  Brownian motion 86
  changes of state 110
  conduction 119
  evaporation 88–9
  gases 84, 85, 86–7, 90–3, 97
  liquids 84, 85, 96
  solids 83, 85, 94
  temperature 85, 87

lasers 145
latent heat of fusion 111
  see also specific latent heat
latent heat of vaporisation 111
  see also specific latent heat
LDRs 196, 198
LEDs 196, 198, 200
length, measurement 2, 3–4
lenses, converging (convex) 141–4
Leslie's cube 120
light
  diffraction 129
  dispersion 145
  monochromatic 145
  properties 127, 131
  reflection 131–3
  refraction 134–8
  total internal reflection 139–40
  white 145
light-dependent resistors (LDRs) 196, 198
light-emitting diodes (LEDs) 196, 198, 200
light rays 131
light-sensitive switches 198
limit of proportionality 35, 36
linearity
  graphs to show 256
  thermometers 99
liquid crystal thermometers 98
liquid-in-glass thermometers 98, 99

liquids
  convection in 118, 120, 1119
  pressure 80–1
  properties 84, 85
  thermal expansion 96
loft insulation 123
logic gates 202–5
logic numbers 202
longitudinal waves 126

magnetic fields
  due to current-carrying coil (solenoid) 210–11
  due to current-carrying wire 209–10
  interaction 212–13
  magnets 155–6
  and moving conductors 212–13
magnetic materials 155
magnetisation, methods of 158–9
magnetism 155–60
  ferromagnetism 157
  induced 157
  test for 157
  theory of 157
magnets 155
  bar 156, 160
  electromagnets 160, 211
  iron used for 160
  permanent 160
  steel used for 160
  temporary 160
magnifying glasses 142–3
manometers, mercury 82
mass 16
  symbol and units 22
  and weight 17–18
mass number (A) 229
melting 110
meniscus 3
mercury barometers 79
mercury manometers 82
micrometer screw gauge 3–4
microwaves 123, 146, 147
mirrors, plane 132
molecular model of matter see kinetic molecular model of matter
moments 38–42
  principle of 39–40
momentum 48–52
  in collisions 61–5
  conservation of 52, 62–5
  definition 48
  formula 48
  and impulse 51
  and Newton's 2nd Law 50
  symbol and units 48
monochromatic light 145
motors, d.c. 214–15

NAND gates 204
neutrons 162, 228
  charge 229
  mass 229
Newton's laws 29, 50
non-renewable energy resources 66
  advantages and disadvantages 70
NOR gates 204
normal 131

265

NOT gates 202, 203
nuclear fission 66, 230
nuclear fusion 67, 230
nucleon number (**A**) 229
nucleus 229
nuclide notation 229, 237

ohmic resistors 177–8
Ohm's Law 177–8
OR gates 202, 203

parachutes 20–1, 32
parallax 2
parallel circuits 187–9, 191
parallelogram rule 46–7
p.d. 171
period
   pendulum swings 4
   waves 126
periscopes 140
permanent magnets 160
physical quantities 242–3
pitch 153
potential difference (p.d.) 171
potential dividers (potentiometers) 194
power 74–5
   electrical 107, 181–5
pressure 76–82
   atmospheric 79
   definition 76
   formula 76
   gases 79, 87, 90–3
   liquids 80–1
   measurement 79–82
   practical examples 76, 78
   symbol and units 76
principal focus 141
principle of moments 39–40
prisms, and total internal reflection 140
proportionality
   direct 35, 256
   graphs to show 256
proton number (**Z**) 229
protons 162, 228
   charge 229
   mass 229
   symbol 229

radiation 118, 119
   absorbers 120–1
   applications 124
   consequences 122
   emitters 121
   reflectors 120–1
radiators, domestic 124
radio waves 127, 147
   diffraction 130
radioactive decay 237–41
radioactive isotopes, uses 236–7
radioactivity 231–41
   background radiation 231, 239
   detection 231–4
   half-life 238–41
   safety precautions 241
   see also alpha (α) particles;
   beta (β) particles;
   gamma (γ) rays

rarefactions 149
ray diagrams
   converging lenses 142–4
   plane mirrors 132–3
real images 141, 143
rectification 197
reflection
   light 131–3
   total internal 139–40
   waves 128
refraction
   light 134–8
   waves 128–9
refractive index 135, 136, 139
relays 196, 199, 211
renewable energy resources 66
   advantages and disadvantages 70–1
resistance 173–80
   definition 177
   effect of heat on 180
   factors affecting 174–6
   filament lamps 180
   determination 179–80
   ohmic resistors 177–80
   Ohm's Law 177–8
   symbol and unit 173
resistors 173
   light-dependent (LDRs) 196, 198
   determination of resistance 179–80
   ohmic 177–8
   in parallel 188-9
   in series 188, 192–3
   variable as transducers 196
resultant forces 28–9, 31
right-hand grip rule 210
ripple tanks 127
Rutherford's scattering experiment 227–8

scalar quantities 7, 46
scientific notation 253–4
series circuits 186–90
significant figures 252
slip rings 220
Snell's Law 135
soft magnetic materials 160
solar energy 68, 70, 71
solidification 110
solids
   conduction in 118, 119
   properties 83, 85
   thermal expansion 94–6
sound
   amplitude 154
   diffraction 129
   echoes 151–2
   frequency range 148
   loudness 154
   pitch 153
   production of 148
   speed 148, 150–2, 153
   waves 126, 148–9, 154
specific heat capacity 104
   determination 106–7
specific latent heat 112
   determination 114–17
spectrum 145
speed 6–8

   definition 6
   from distance–time graphs 8-9
   and velocity 7
speed–time graphs 9–12
split ring commutators 214, 215
spring constant 36
stable equilibrium 45
standard form 253–4
static (electrostatic) charges 161, 164–5
step-down transformers 223
step-up transformers 223
stiffness of spring 36
stopwatches 5
symbols 242–3

television (TV) reception 130
television (TV) sets 181
temperature
   measurement see thermometers
   and motion of molecules 85, 87, 93
temperature-operated alarms 199
temporary magnets 160
terminal velocity 19, 20
thermal capacity 104–5
thermal expansion 94–7
   applications 95
   consequences 95–6
   gases 97
   liquids 96
   solids 94–6
thermistors 196, 199, 200
thermocouples 102–3
thermometers 98–103
   calibration 98, 100–1, 103
   constant-volume gas 99
   fixed points 98
   linearity 99
   liquid crystal 98
   liquid-in-glass 98, 99
   range 99
   sensitivity 100
   thermocouples 102–3
tidal energy 69, 71
time
   measurement 4–5
   symbol and units 6
total internal reflection 139–40
transducers 195–6
transformers 222–6
   efficiency 224–5
   step-down 223
   step-up 223
   and transmission of electricity 225–6
transverse waves 127
truth tables 202
   AND gates 203
   combinations of gates 205
   NAND gates 204
   NOR gates 204
   NOT gates 203
   OR gates 203
   NOT gates 203
   OR gates 203

ultrasound 148
ultraviolet rays 147

units 242–3
unstable equilibrium 45

vacuum flasks 124
vector quantities 7, 46–7
   resultant of 46–7
velocity 7
   and speed 7
   terminal 19, 20
virtual images 132, 143
visible light 147
volt 171
voltage (drop) 171
voltmeters 171–2
volume
   measurement 3, 25, 27
   symbol and units 22

wall insulation 123
water waves 127–9
wave energy 68, 71
wave formula (equation) 126
wavefronts 127
wavelength 125, 153
waves 125–30
   amplitude 125, 154
   diffraction 129–30
   electromagnetic 127, 146–7
   frequency 125, 153
   longitudinal 126
   period 126
   reflection 128
   refraction 128–9
   sound see sound
   speed 126
   transverse 127
   water 127–9
   wavelength 125, 153
   see also light
weight 16
   and mass 17–18
white light 145
wind energy 68, 70, 71
work (done) 73

X-rays 147